Reimagining Time

Reimagining Time

A light-speed tour of
Einstein's theory of relativity

Tanya Bub and Jeffrey Bub

Yale

UNIVERSITY PRESS

New Haven and London

Yale University Press books may be purchased in quantity for educational, business, or promotional use. For information, please e-mail sales.press@yale.edu (U.S. office) or sales@yaleup.co.uk (U.K. office).

Printed in the United States of America.

Library of Congress Control Number: 2020943781

ISBN 978-0-300-25012-1 (hardcover : alk. paper)

A catalogue record for this book is available from the British Library.

This paper meets the requirements of ANSI/NISO Z39.48-1992 (Permanence of Paper).

10 9 8 7 6 5 4 3 2 1

For Einstein, who liked trains

CONTENTS

PREFACE

This book is the result of a bet I surprisingly won. It started with a graphic novel project I was supposed to do with my dad, destined to die on the drawing board, and a long-frustrated desire to understand Einstein's theory of relativity. To clarify, it's the graphic novel that died. My dad is fine.

Regarding the long-frustrated desire to understand relativity, here's what I wanted: to really get what it means to say time is relative; to see why energy and mass are equivalent; to grasp how it is that nothing can move faster than light. That kind of thing.

Here's what I didn't want: to plug numbers into equations I didn't really understand, thereby turning them into other numbers I didn't really understand.

Like many people with such long-frustrated desires, I have a collection of books on relativity that I started reading but never finished. Since you have this book in your hands, you might possibly know what I'm talking about.

Now flip back in time to 2018. My dad, Jeffrey Bub (a theoretical physicist who has spent a good part of his life thinking about problems in the foundations of physics), and I are publishing a graphic novel on quantum mechanics called *Totally Random*. On the high of getting the first book off to press we decide to do a sequel. The topic? Einstein's theory of special relativity. While this may seem an odd thing for someone with the aforementioned long-frustrated desire to take on, my dad reassures me he's got this (he has lots of experience plugging numbers into equations he does understand), so we agree to give it a try. We get together. He earnestly attempts to further my understanding by having me plug numbers into equations. We part company agreeing to disagree on the direction of the book for the time being.

After a month or two of more or less flailing around on the project, I come to a rather surprising realization: I've never actually read anything about relativity written by Einstein—the person who came up with the theory in the first place—that wasn't quoted by somebody else. It occurs to me maybe I should do something about that. So, I Google "Einstein's original paper on relativity" and get this:

---

hermes.ffn.ub.es › luisnavarro › Einstein_1905_relativity PDF

ON THE ELECTRODYNAMICS OF MOVING BODIES

of which will hereafter be called the "Principle of Relativity") to the status … †Editor's note: In Einstein's original paper, the symbols (H, Z) for the co-ordinates of …

by **A EINSTEIN - 1905** - Cited by 1357 - Related articles

Not terribly inviting. Regardless, I click on the link and half-heartedly skim the first few pages until I hit these words:

> two systems of co-ordinates
> in uniform translatory motion.

An image pops into my head. Trains.

Two trains, side by side,
moving in opposite directions.

Einstein was describing a scenario I could picture. I printed the paper and doodled two trains in the margin across from those words, very much like the trains you'll soon be riding as you make your way through this book. I was to spend the better part of the next two years satisfying my long-frustrated desire to understand special relativity by painstakingly taking Einstein's words quite literally and playing them out on imaginary trains.

So where did the bet come in? After discovering the "trains" in Einstein's paper I called my dad and breathlessly told him about them. He was not impressed. He informed me that Einstein's original paper was the old way of looking at relativity and that there were new ways that were much better. Sulkily, I told him the new ways weren't better for me. So, he flew across the continent (in a plane) from Washington, DC, to Victoria, BC (where I live), to see my trains, which I'd cut out of cardboard for demonstration purposes (true story). After much sliding around of pretend trains using a bead to indicate the position of light, he sort of got what I was on about and eventually agreed that since we were working on a graphic novel (the one that was destined to die on the drawing board), we could do trains if I wanted because really, how many equations could the average comic-book reader stomach anyway.

Then he said this (paraphrased): Your trains may give readers a qualitative feel for relativity but they won't deliver a quantitative understanding.

In other words, your trains may convey some of the concepts but they can't produce numbers. For numbers you need equations.

And I said (paraphrased): I bet you're wrong.

Arguably a somewhat silly bet to make.

However, I did have a glimmer of evidence to suggest I could be right. Even by that point, very early on in the project, I could see that while there were clearly many wrong ways to represent light on moving trains, there could logically only be one right way. That right way would necessarily exemplify the right numbers. The trick was finding it.

The problem was length contraction. Einstein describes how lengths contract in the direction of motion by means of equations. Without length contraction my trains were doomed to be yet another bad half-wrong relativity metaphor with little more to offer than the sad flying ball that gets flattened into a pancake for no apparent reason (Google "length contraction" to see said ball). The question was—could the length contraction factor be gleaned without equations? Spoiler alert. It can. You'll see exactly how in Episode III.

At this point we sent the book in its graphic

novel incarnation to our agent, Peter Tallack, who unceremoniously informed us that although he liked the concept, the market for challenging science comics had more or less dried up. So, the graphic novel died on the drawing board that day, only to be resurrected in the form of the slightly more conventional illustrated text you are currently reading.

By now, despite the fact that even Jeff was fully on board with the trains, having seen them convey not only relativistic concepts but also relativistic values, neither of us was at all certain we could deliver the big prize without resorting to math. I'm speaking, of course, of the ultimate equation, Einstein's ubiquitous $E = mc^2$, the culmination of, and arguably most famous part of, the theory of special relativity. To our mutual astonishment it turned out that the specific relationship between energy and mass was in fact right there on the trains all along. All that was needed was a touch of momentum. You'll get to play it out for yourself in Episode IV.

While there are no equations in the book at all (the toughest math you'll have to do is double or halve a number), reading it requires focus and concentration. The reason this book is challenging, in spite of its slender profile, pretty pictures, and chatty repartee, is that understanding relativity requires you to change very fundamentally the way you think about space, time, and matter. And that is a much harder thing to do than plugging numbers into equations.

For readers already familiar with relativity, here's a little tip: while length contraction isn't explicitly addressed until the second half of the book, rest assured that the examples in the first half do take it into account. If you read carefully, you will note that the first two scenarios are described in terms of approximations (about half, around midway, etc.). The intention being to deliver the qualitative goods (time and simultaneity are relative!!!), without bogging readers down with the quantitative stuff until they're ready to see where it comes from for themselves. If you do feel the need to anchor the examples in specifics as you go along, flip back and forth between the chapters and the corresponding Geek Notes at the end of the book, where you'll find the exact values and equations that describe each scenario (courtesy of Jeff).

You may also notice I steer clear of certain conventional ways of talking about relativistic effects, like time slowing, for example. The intention is to keep things very tangible by only drawing conclusions directly apparent to the reader, such as the measurable amount of time and space between particular events.

It's probably worth mentioning that the trains in this book never accelerate or decelerate. That's because Einstein's theory of special relativity only deals with constant relative motion. This means, regardless of which train you're on, it feels perfectly stationary. There's no wind, or sensation of movement at all.

Think of your train as parked, with the other train passing by alongside yours.

Also, I'd like to express my gratitude to Michel Janssen and John Norton, who very generously read drafts of the book, providing invaluable feedback. Any errors or missteps are of course my own.

And finally, thanks to my dad, Jeffrey Bub, co-author and devil's advocate extraordinaire, without whom this book (and especially the equation-laden Geek Notes at the back) would never and could never have been written.

Introduction

LIGHT GOES AT ONE SPEED

There are some things everyone should understand before they die. For example:

NOTHING WORTH HAVING COMES EASILY

A BIG DESTINATION TAKES A BIG JOURNEY

TRUE UNDERSTANDING IS BORN OF EXPERIENCE

Universal things like that. Here's another one:

SPACE AND TIME ARE RELATIVE

You may live in New York, or Lisbon or Harar, but regardless of postal code every one of us lives out a life in space which necessarily unfolds over time, so really shouldn't everyone know that space and time aren't what they seem?

Space isn't a rigid grid-like backdrop and time isn't a universal metronome ticking away behind the scenes. Space and time, as you will soon see, are way more interesting than our day-to-day life experiences lead us to believe; in fact, they stretch and shrink like a fantastical cosmic Slinky in a funfair house of mirrors.

The first person to notice this was a rumpled 26-year-old, third-class patent clerk working out of an office in Bern, Switzerland. This unlikely character would go on to become one of the most famous people in the world. You've no doubt heard of him. His name is Albert Einstein.

With his fabulous disheveled hair, tendency not to wear socks, and notoriously messy desks, Einstein has since become the poster boy for genius, recognizable instantly around the world. But before all that, there was a time when he was an anonymous clerk who had the distinction of being the only person to have glimpsed a mind-blowing truth about the universe. This book is about what that clerk saw, Einstein's biggest idea: a game-changer that came to be known as the theory of relativity.

Many people are surprised to learn that Einstein didn't discover relativity by doing experiments. He also didn't have access to any information that wasn't already well known to the scientific community at the time. But Einstein did do something no one else was willing or able to do: give up some very strong, very human, and, as it turns out, very wrong ideas about how space and time work. Putting aside common-sense beliefs, Einstein was able to see the structure of the universe in an entirely new way.

Even though Einstein published his eye-popping insights over 100 years ago, most people don't really get what he was talking about. This isn't very surprising. Understanding relativity means resisting the pull of a lifetime of experience, all of which supports the wrong idea that we exist in a rigid space for a fixed period on a single timeline over which all events unfold in a universal order.

That's where this book comes in. It's a prop designed to give you a different experience.

One that will help you push back against what you think you know about space and time.

So, what do these common-sense-but-wrong intuitions look like? They're not hard to find, because they form the foundation of all your ideas about the physical world. They are strong and seductive and hard to resist.

Want to feel their pull? Then let's talk about baseball.

Not Incredible

FASTEST BASEBALL PITCHES OF ALL TIME

Aroldis Chapman	106 mph
Joel Zumaya	104.8 mph
Mark Wohlers	103.5 mph
Jonathan Broxton	102 mph

When you look at these stats, you may well find them to be incredible. If so, what's likely to have impressed you is that a human being is capable of throwing a baseball at speeds upwards of 100 miles per hour. However, you're unlikely to be amazed by the fact that baseballs move at different speeds. That's just run-of-the-mill stuff for ordinary objects like baseballs.

You probably also won't be surprised to know that if Aroldis Chapman were to make his 106-mile-per-hour pitch while standing on a train going at 100 miles per hour, the top speed of that pitch would register as 206 miles per hour when measured by a radar on the ground. That's because from the ground the speed of the ball gets added to the speed of the train.

BALL SPEED MEASURED FROM TRAIN	BALL SPEED MEASURED FROM GROUND
106 mph	206 mph

That's what we expect moving objects to do.

So, not incredible.

Incredible (Literally)

On the other hand, you probably would find it literally incredible if that same pitch, the one Aroldis threw from the train, clocked a maximum speed of 106 miles per hour on both the radar on the train *and* the radar on the ground, even though the train is moving at 100 miles per hour relative to the ground. You might not even be able to get your head around it. Maybe you'd think there was something wrong with the radars because, let's be honest, these stats don't make sense considering how we think objects move through space in time.

BALL SPEED MEASURED FROM TRAIN	BALL SPEED MEASURED FROM GROUND
106 mph	106 mph

Yet that's how light works. Really.

Unlike baseballs, light goes at one speed. This speed happens to be about 186,000 miles per second in empty space. This means that if instead of a ball we give Aroldis a laser gun and have him fire off a light pulse, that pulse will clock in at 186,000 miles per second. If he fires that same laser from a train going 100 miles per hour, the light pulse will still clock in at 186,000 miles per second, amazingly, from both the train and the ground. This is true if Aroldis fires the laser from a train going at 100 miles per hour or 100,000 miles per second. The light pulse's speed will be correctly measured as 186,000 miles per second, measured from both the train and the ground. Really.

LIGHT SPEED MEASURED FROM TRAIN	LIGHT SPEED MEASURED FROM GROUND
186,000 mps	186,000 mps

This fact is hard to swallow: the speed of the train doesn't get added to the speed of the light. That's what it means to say light goes at one speed; Aroldis simply can't make light go any faster (or slower) by firing the laser from a moving train.

You will soon see that this one weird fact—light goes at one speed—wreaks havoc with our very strong and very wrong ideas about how space and time work. They don't work the way you might think they do.

How do we know light goes at one speed anyway? Light goes so fast that for a long time its speed was considered infinite. Infinitely fast light would take no time to travel any distance, near or far. The first person to prove this wrong was Danish astronomer Ole Roemer. In 1676 he noticed that eclipses of one of Jupiter's moons took place sooner than predicted when the earth was closer to Jupiter and later when the earth was farther away. He reasoned this was due to the increased distance the light had to travel to reach the earth. Light that takes longer to travel a greater distance must have a finite velocity. Taking into account the relative distance of the earth and the resulting time lag, Roemer figured the speed of light had to be about 133,000 miles per second—impressively close for 1676! By 1728 science had progressed to the point that English physicist James Bradley calculated the speed of light within 1% of the true value.

Light having a speed raised a new question: Relative to what? Light could pass through air but also through the vacuum of space. Logically light's speed should be measured against the backdrop of whatever the light was traveling through. In the late 19th century that stuff, the hypothetical medium through which light supposedly propagated, was called, quite poetically, luminiferous aether. Two scientists, Albert Michelson and Edward Morley, reasoned that the earth, in constant motion as a result of its orbit, must be moving through the stuff too. It stood to reason that light's speed measured on earth should vary depending on the earth's velocity and direction of motion through light's medium. So, they devised a clever experiment in which they aimed beams of light in different directions, expecting a discrepancy in their velocities to result from the relative motion of the earth through the aether. But there was no discrepancy. None whatsoever. The speed of light was always exactly the same. Michelson and Morley found this about as believable as a baseball clocking in at the same speed on a radar on the ground and a radar on a moving train. It just made no sense. So they repeated the experiment many times, increasing their instrument's precision and sensitivity in an effort to get a result they could swallow. No difference. The result remained stubbornly negative. Light evidently had one speed only. The Michelson-Morley experiment came to be known as the most famous failed experiment in history. Because instead of revealing the motion of the earth through light's medium, they exposed a much more interesting fact: such a medium doesn't exist at all!

With no medium against which to measure the speed of light, light had a velocity of 186,000 miles per second relative to… WHAT? Anything? Everything? Very perplexing!

Einstein spent a lot of time thinking about this. He wasn't especially concerned with the actual speed of light per se or even the details of the aforementioned experiments for that matter. What he was interested in was the very idea of one-speed light and what its existence had to say about the universe. So he started with that and took clear, logical steps from there. These steps led him to a completely new understanding of the nature of space, time, and matter.

Here's an example of the kind of reasoning he used.

LIGHT GOES AT ONE SPEED SO…
Light Can't Overtake Light

Reason: one thing has to be going faster than another to pass it. Since light goes at one speed, one light pulse will never pass another.

By methodically and fearlessly exploring the meaning of one-speed light Einstein was able to reveal astounding truths about the universe that no one had noticed before. Things like…

TIME IS RELATIVE
LENGTHS GET SHORTER WITH MOTION
THE UNIVERSE HAS A SPEED LIMIT
ENERGY & MASS ARE INTERCHANGEABLE

Einstein reasoned using math. But there's no math required for this book. The work done by the math is instead done by pictures. The pictures show what Einstein proved mathematically, so that you can see the things he saw without having to do any calculations. This works because the math and the pictures both describe the same thing—a world in which light goes at one speed.

Some of the pictures are here in the book, but others will take place in your mind. The mental pictures are really more like movies. This is not a gimmick. Imagining how one-speed light moves through space will allow you to grasp what is usually described by mathematical equations that calculate change over time. In any case, the movies are not only enlightening, they're also kind of fun.

Some pictures will give you a feeling for relativity; others will give you facts. If you're the sort of person who likes numbers, you'll be happy to learn that you can get those from the pictures too. For example, you might want to know exactly how much time passes for two people moving relative to each other at a particular speed, or maybe you're interested in the actual fraction by which an object in motion contracts. By the end of the book you'll find that the pictures reveal all sorts of meaningful relativistic details that match up perfectly with the ones calculated in the usual way, that the values are all baked right into the very idea of one-speed light.

Finally, you'll understand how two seemingly unlike

things, energy and mass, ended up on opposite sides of the world's most famous equation, $E = mc^2$, and precisely why a very large, very specific amount of energy is equated with a very small, very particular quantity of mass and what the speed of light has to do with it.

And all this, amazingly, will come from a careful unpacking of one little fact:

Light goes at one speed.

Here's what you need for the ride:

A Good Imagination & Some Trust

You'll soon be asked to do odd things like "watch" a light pulse move in slow motion, or jump onto a train moving at half the speed of light. This is clearly not possible in real life. Try not to worry about that kind of thing too much. Also, if you've already done some reading on relativity, that's great, but put what you've learned aside for now. You won't be needing it for this particular trip. Should you find yourself getting hung up on details, please check out the Geek Notes at the back of the book. They're chock-full of reassuring behind-the-scenes numbers.

On the subject of numbers, you may notice that the frame rate in the visualized shorts has been customized for your viewing pleasure. A *Reimagining Time* second is really about 100 nanoseconds, which is more or less how long light takes to go from one end of a train car to the other.

After reading the next few pages you'll see why one-speed light means time has to be relative. This revelation about the true nature of time usually happens at the other end of a calculator and, even then, generally only for people who glean meaning from mathematical equations. Here, it will come to you in a specially designed "room" furnished with props concocted to help you visualize the movement of one-speed light. Since the room will be constructed by you, in your mind, you'll be able to slow down the action, pan in and out, replay the scene from different perspectives, and discover things that are not at all obvious in our day-to-day world.

Once in the room you'll make a series of simple observations that follow a line of reasoning first conceived by our 26-year-old patent clerk, Albert Einstein, which led to him to discover the relativistic nature of the universe in 1905 simply by thinking about light.

If you are ready, close your eyes for a moment, take a deep breath, and turn the page.

Episode I

LIGHT GOES AT ONE SPEED
SO TIME HAS TO BE RELATIVE

When you open your eyes, you find yourself lying on your back looking straight up into empty space. You scan left and right, but there's nothing to see. The surface you're lying on is flat and smooth. As you sit up you see that you're on a white flatbed freight train, with no seats, no ceiling, no walls. You peer over the edge. The train doesn't seem to be standing on anything. It's just sort of floating there in nothingness. A featureless freight train, made up of one identical white train car after another as far as your eye can see in either direction.

You get up and start walking.

After some time, you spot something. A welcome diversion, just a few train cars up. As you get closer you see it's some kind of a gun; a laser, in fact.

You pick it up.

The laser has a built-in digital timer on the side that reads 0:00 seconds and next to it, in fine print: **LIGHT GOES AT ONE SPEED**

One speed, huh?

As you wonder "But what speed?" a glint at the other end of the train car catches your eye. It's a small mirror mounted to a signpost at shoulder height. Could you hit it from all the way across the train car?

You close one eye, slowly raise your arm, and, pointing the gun straight at the mirror, pull the trigger.

Bzzzp! A light pulse shoots out of the gun and the gun's timer starts ticking.

Your aim is true. The light pulse hits the mirror and reflects straight back at you! You duck just in time. The light pulse whizzes past your head. At that moment you release the trigger and the timer stops.

You glance at the gun.

It reads 2:00 seconds.

REFLECT

2:00

IT GOES AT ONE SPEED

Well, that answers your question, anyway. Light goes at one speed. Since you were standing at one end of the train car when you fired at the mirror at the other end and the one-speed light pulse took two seconds to bounce back to you, it must have taken one second to cross the train car each way—one second there and one back. Light's speed is, then, the length of the train car per second.

But are all the other train cars the same length as this one? It occurs to you that light's one speed could be used to check. If one-speed light takes the same amount of time, one second, to cross the other train cars too, they're the same length.

You decide to find out. You walk further along the train so you are now two train cars away from the mirror. You aim and fire—the timer starts—the light hits the mirror, reflecting back—again you duck, glancing at the timer as the pulse whizzes by: four seconds passed since you pulled the trigger. Two seconds there and two back, to go the length of two train cars. That's one train car per second. Both train cars are the same length.

You try again, with three, four, five, six train cars, until you are satisfied. The length of each train car is the distance one-speed light goes in one second. One-speed light's travel time is a nice way to measure distance!

Now pan out so that you see yourself from far away, standing at one end of a train car, your arm extended, pointing your gun, along the length of the train. You fire! Watch the light pulse emerge from the gun's barrel in slow motion as the timer starts ticking. See it cross the train car as you hear the tic-tic-tic of the timer. After one quarter of a second the timer reads 0:25 and the pulse is a quarter of the way across the train car ... tic-tic-tic ... after half a second the pulse is halfway and the timer reads 0:50 ... tic-tic-tic ... When the timer reads 1:00, the light pulse passes the opposite end of the train car and keeps going. The pulse reaches the end of the second train car as the timer ticks off 2:00 seconds, and it crosses the end of the third train car as it ticks off three seconds, the passage of light marking the passage of time, one second per train car, train car after train car.

That's when it hits you. You don't even need the timer anymore! You can see how much time has passed since the laser was fired just by looking at how far the light pulse has gone along the train. Uniform cars plus one-speed light make the train a perfect clock—the position of the light pulse on the train measures the passage of time since the trigger was pulled.

You fire again and this time freeze frame when the light pulse is exactly 2.5 train cars away. You know before checking the timer that 2.5 seconds have passed since you pulled the trigger. You fire again and this time freeze frame when the light is 5 train cars away. Five seconds have passed. One-speed light's travel distance is a nice way to measure time!

While experimenting with your gun you notice something coming towards you.

It's approaching frighteningly fast, getting larger and larger by the moment until—WHOOSH!!—it passes you. It appears to be another freight train very much like the one you're standing on except this one is black. It also seems to be endless and is moving at an alarming speed with a gutter of just a few inches between it and your white train.

What would it feel like to go that fast? You have an almost overwhelming desire to experience that speed. A crazy idea comes to you. Could you jump from your white train onto the fast-moving black one?

Once you think of it you can't get the idea out of your head. You hold on tight to your gun, grab the little mirror, brace yourself, and leap!

Your foot makes contact with the black surface of the fast-moving train. You tumble and roll for several train cars, eventually coming to a complete stop. You dust off and stand up. So this is what it's like to ride the speeding black train. It's not very exciting. In fact, the black train doesn't seem to be moving at all. What's more, from the black train, the white train is the one that's zooming enticingly by at an outrageous speed. Which train is actually moving? Since the trains are in empty space there's no context for motion except for the other train. From the white train the black one is moving. From the black train the white one is moving. The motion of each is relative only to the other. When you're on either train, black or white, it feels still, with the other train zipping by alongside it.

Since you're here, you decide to look around. The black freight train is also made up of a series of seemingly identical train cars. You guess they're the same length as the ones on the white train, but how can you be sure? Then you remember the light gun. You set up your little mirror, go to the opposite end of the train car, aim, and pull the trigger. It turns out you're right. The timer tells you that it takes the light pulse two seconds to reflect back at you, meaning it crossed the length of a black train car in exactly one second, just as it did on the white train. You take a few more shots, first in one direction and then in the other, placing the mirror at various distances away until you are certain. Light goes at one speed and takes one second to cross a train car, black or white.

You're feeling good. You own this thought experiment.

You're at home with one-speed light. It just makes so much sense. You clearly see that one-speed light is the gold standard for measuring train-car lengths and that uniform train cars plus one-speed light make a perfect clock. You're so confident on this point that you barely even look at the timer anymore. The reason these facts feel so solid, so indisputable, is that they're just a natural consequence of the fact that light goes at one speed. One-speed light means it could be no other way. You wonder what else one-speed light has to say about the world. You sit down to think about it.

Here's something. Light can't overtake light. You vaguely remember reading that somewhere. Anyway, it must be true because one thing has to be going faster than another to pass it. Since light goes at one speed, one light pulse can't go faster than another. That seems pretty obvious.

What would light not overtaking light look like? How would you even test that anyway? What if lasers were fired from the white train and the black train? Since light can't overtake light, one pulse could never pass the other. And what if the shots were timed perfectly, with each gun on either train firing in the same direction, at the very same time, at the instant the guns passed each other? One-speed light pulses fired together would stay together. You imagine two light pulses emerging in tandem from two guns, one fired from the black train, the other from the white, remaining side by side, as they move along the lengths of the two trains, always neck and neck, neither able to pass the other because light goes at one speed.

When you come to think of it, you realize that, unlike ordinary bullets, which can go at different speeds, the speed of a light pulse is unaffected by the speed of the gun it's fired from. A light pulse fired from a gun on the moving train goes the same speed as a light pulse fired from the stationary train. Naturally. It has to be this way because light goes at one speed. And light takes one second to cross a train car.

Light pulses fired together stay together. It sounds innocent enough. Why, then, does it leave you feeling uneasy? If only you could see it for yourself. Then you could put your finger on exactly what's bothering you. Quite suddenly you feel an irrepressible urge to play out the double-shot scene you just imagined. And why not? This is your thought experiment. No one can stop you from doing whatever you like!

Reaching deep into your imagination you split yourself into two observers, each with an identical laser gun. One you is now on the black train, the other you transported far down the white train. You find you can flip back and forth between the two perspectives at will, a bit like cutting between movie scenes—each train taking its turn being the stationary one with the other train in motion—until you are quite used to it. Then you remember why you're doing this in the first place. You're here to see two light pulses fired together stay together. You dutifully switch your point of view to the black train and get down to business.

You are now on the stationary black train. You lean slightly into the narrow gutter between the two trains, bracing yourself because the white train is coming at you very quickly, each train car approaching and whizzing by, one after another after another.

Then you see what you're waiting for. The other you riding the white train is straight ahead, still far in the distance, but rapidly closing in. The white-train you, whose back is to you so that you are both facing the same direction, is clearly also preparing for the moment when the two guns cross paths. You steady your arm, point your laser right along the gutter between the two trains, and wait. It's nearly time for action! Momentarily, the two guns, facing the same way, align—this is it—Bzzzp! **Freeze frame!**

Perfect timing! Both the white-train you and the black-train you pull your triggers in the instant of passing.

Resume the action
in slow motion.

Watch the two light pulses,
inches apart, locked together
at one speed, moving along the gutter
between the black train and the white train.

After a quarter of a second the pulses
are a quarter of the way across your black
train car. In that time the white train has also
moved, the other you receding behind your back,
even as the approaching far end of the white train
car draws closer.

About half a second after you both fired your guns, the two light pulses, now about half-way across your black train car, have just reached the white train car's advancing edge. (See Geek Notes for exact numbers.) **Freeze frame!**

You look at the position of the light on your black train car and see that it took the light pulses about half a second to cross the length of the white-train you's white train car.

Now rewind the scene to experience the same double shot from the white train.

This time you're on the stationary white train with the black train whizzing by. You look back over your shoulder and wait for the black-train you to approach from behind. There! You brace yourself as the moment comes. The two guns line up and in that instant you both fire—Bzzzp! The action now plays out in cinematic slow motion. You watch the light pulses going in the same direction as the passing black train. After one second they reach the opposite end of your white train car. **Freeze frame!**

You examine the frozen scene before you. In the time that passed since you both fired, the black-train you has moved and is now at your train car's midpoint. The far end of the black train car is now well beyond the end of your white one.

In one second the light pulses have gone the length of your white train car, and the black-train you, now halfway across your train car, has gone half that distance. The position of the black-train you tells you that the black train is moving at half a white train car per second and is therefore going at half the speed of light. Good to know.

The light pulses, now at the far end of your white train car, are only partway along the moving black train car's length.

They will eventually catch up to the black train car's receding end, but it will take time because they're chasing a moving target. Right now, they're somewhere around the black train car's midpoint, meaning it took the light one second to go about half a black train car length.

You stare at the position of the light pulses. How much time has passed since you pulled the trigger? Uniform car lengths plus one-speed light make each train a perfect clock. To see how much time has passed you need only look at how far the light has gone. It couldn't be simpler. But how much time has passed since the shots were fired? The one-speed light pulses are about halfway along the length of the black train car. So, about half a second has passed since the shots were fired. And the very same light pulses are also at the far end of the white train car. So one second has passed since the shots were fired. Which perfect clock is telling the right time? How could you possibly pick one over the other? You gasp as the answer comes into focus. Both clocks are correct!

Clearly, the light pulses can't cover the same distance on two train cars moving relative to each other. You know that and yet somehow this still shocks you. Your heart quickens as you consider how much time one-speed light takes to cross any train car. You don't even need to look at the timer on the white train's laser gun to know it reads 1:00 second as it always does and as it must, because it takes one-speed light one second to cross one train car. Period. One second has passed on the white train.

The position of the light on the black train car is equally telling. The timer on the black train's laser gun can only indicate that about half a second has passed. It's not negotiable. There's no way around it. One second has passed on the white train between when the guns were fired and the light pulses reached the end of the white train car—and as you can clearly see by the position of the light pulses on the black train, those events occurred in about half a black-train second.

Uniform car lengths plus one-speed light do make the trains perfect clocks. It's just that the clocks, and time itself, are a lot more interesting than you thought!

You let it slowly sink in. Time passes at different rates depending on motion. One-speed light has made that perfectly clear. Fleetingly you consider that in the time between when the guns were fired and the light pulses crossed the far end of the white train car, the white-train you aged a full second, while the black-train you aged only about half that.

The thought of it all makes you slightly dizzy. Why did no one tell you time worked this way? You suppose maybe they did. So this is what it means to say time is relative. Really, though, it's one of those things you need to experience for yourself.

The action now resumes, both light pulses continuing on their way at the only speed they know.

And just as you think you're starting to get your head around the idea of relative time, a nagging little thought pops into your head. In the one second it took the light to travel the length of the white train car, about half a second passed on the black train. But isn't the black train/white train setup symmetrical? Why then should less time pass on the black train than on the white train? What if the light guns had instead been fired together in the opposite direction?

You imagine it: this time the light is shooting away from the rapidly receding black-train you. In this case, the pulses pass over more than one black train car by the time they reach the end of the white one. So more than a second would have passed on the black train between events separated by one white- train second. How can more *and* less than a second possibly pass on the black train in the one second it takes light to travel the length of a white train car?!?

Your eyes snap open.

End scene.

...until at last it came to me that time was suspect!

—Albert Einstein

Light goes at one speed, so time has to be relative. Time does not tick away at a single set rate to the beat of a universal drum. Rather, the time between events depends on motion, a stunning truth exposed by one-speed light.

If you find this surprising feature of time hard to believe, you're not alone. Einstein himself wrestled with the seemingly contradictory nature of light from when he was a boy of sixteen until finally, in a flash of insight, the solution came to him: there is no universal time! This epiphany, and Einstein's audacious willingness to reimagine time itself, was to open the door to the slew of mind-blowing discoveries revealed in his revolutionary 1905 paper on relativity.

So how would Einstein explain the fact that more and less time passes on the black train between events that occur after one white-train second? With a concept he dubbed the relativity of simultaneity. Read on to see how this bizarre feature of relativity accounts for such apparent paradoxes of one-speed light and brings length contraction into the picture.

Mission Impossible?

Imagine you have a gun that shoots two one-speed light pulses in opposite directions.

<<—<<—<<— LIGHT GOES AT ONE SPEED —>>—>>—>>

And you get this apparently impossible mission:
You must fire your gun so that you
hit Alice before Bob
and hit Bob before Alice
and hit both simultaneously
ALL WITH ONE SHOT.
Could you do it?

The next episode is all about
that one shot and what it has
to say about the world we live in.

ALICE   BOB

You will also see how length contraction, a seemingly paradoxical aspect of Einstein's theory of relativity, comes into the picture. In a nutshell, you will observe the white train as shorter than the black train and the black train as shorter than the white train.

These somewhat surprising facts about space and time will be exposed by your two-sided gun's light pulses, which behave very differently from ordinary bullets in one critically important way: ordinary bullets go at different speeds. Light doesn't.

That is the key to seeing what Einstein saw.
Light goes at one speed.

You may think that keeping this point in mind will be easy. Sure, it's easy when you don't have to give anything up for it. The going gets tough, however, when ideas about space and time, grounded in a lifetime of experience, are at stake.

If you're like most of us you'll read this chapter and at the beginning feel quite satisfied that you understand what's going on. Then suddenly you may find yourself on a white train watching the bullets fly and think to yourself, "Something tricky has happened," or "I've missed something," or maybe "This book doesn't make any sense after all." It hasn't, you probably didn't, and it does.

What's more likely to have happened is you've been asked to surrender a firmly held but wrong belief in exchange for one-speed light. And that's a hard trade to make.

The next few pages are here to help you pinpoint where exactly the weirdness comes in, to combat the possible onset of doubt or confusion that may occur when you read the following episode. The weirdness results from the fact that one-speed light cannot be made to go any faster (or slower) when fired from a moving gun. Here's how Einstein put it:

Light is always propagated in empty space with a definite velocity c which is independent of the state of motion of the emitting body.

So, what does that mean? Imagine your double-barreled light gun at the very center of these two pages. Press the button on top to fire the gun. Go ahead. Bzzzp! The light pulses emerge from both sides at the same time, and in tandem go in opposite directions, always the same distance from where they were fired (because light goes at one speed), until … they both reach the left and right edges of the book simultaneously. Makes sense, right? One-speed light means it couldn't be any other way.

But what if the gun is moving when it's fired? This time imagine your gun in motion, coming in from the left-hand side of the book and traveling towards the center. Prepare yourself, because at the moment it reaches the middle, you will again press the button and fire. It's nearly there, get ready, get ready, and… FIRE! Bzzzp! Perfect timing! Watch the two light pulses emerge from the gun and make their way, in tandem, both always the exact same distance from the firing point at the center of the book, even as the gun continues on its path towards the right-hand side of the page. Like last time, both light pulses reach the edges of the book simultaneously.

Notice that the fact that the gun was moving when it was fired didn't make one whit of difference. Moving, not moving, it doesn't matter. The motion of the light is measured from its firing point regardless of where the gun happens to be before or after being fired, because unlike ordinary bullets, one-speed light simply can't be made to go any faster or slower (even by a moving gun). Read Einstein's quote on the opposite page for reassurance, as necessary.

ONE SPEED ·->·->·->·->·->·->

This means light pulses fired from our double-barrelled gun will always be the same distance from where the gun was fired, because for one light pulse to be further away than the other, it would have to go faster, and one-speed light pulses don't go faster or slower. As you'll soon see, this little fact checkmates the very notion of universal time.

Let's replay the first scenario from a different perspective. This time the gun is in its fixed position in the middle of the book, but the whole book is moving relative to you. The easiest way to imagine it is to pretend your left thumb is you. Make the universal sign for thumbs-up directly in front of your face. Go ahead; no one's looking. That thumb is now you. Next, hold the open book up with your right hand, and slowly move it from right to left behind your thumb. The moment thumb-you and the gun at the center of the book line up—Bzzzp!—it fires, and the book keeps going.

See the two light pulses emerge from both sides of the gun in tandem, moving so that they are always the exact same distance from your thumb, which is where the gun was when it fired. Remember, the fact that the gun is in motion has no impact on the speed of light, since light always goes at the same speed. All the while the right side of the book draws nearer and the left side retreats, until the light pulse fired to the right hits the advancing right edge of the book first. First!

FIRE!

THE LIGHT PULSE HITS ADVANCING RIGHT-HAND SIDE OF THE BOOK FIRST!

Which means we have a single scenario where light pulses fired from a gun in the center of the book hit both edges of the book simultaneously (as you saw when you played out the scene the first time), AND also hit one side before the other (the very same shot played out with the book moving relative to you).

In a nutshell: Light has one speed. A light pulse goes at that speed for all observers, even if they are moving relative to one another. And if I'm moving, say at half the speed of light relative to you, and we both correctly measure one-and-the-same light pulse as going at light speed relative to ourselves, then my time and your time simply can't pass at the same rate, and events that happen simultaneously for you, like light pulses reaching opposite edges of the book, will not occur at the same time for me. It took Einstein to point out that one-speed light is just not compatible with universal time or absolute simultaneity.

The next episode has no math or numbers whatsoever. Instead you'll get to the punchlines by playing out a single shot that accomplishes all three conditions of Mission Impossible in a specially designed theater of the imagination. The setup? Two trains moving at half the speed of light relative to each other, where you'll witness the same two events occur at the same time, then in one order, and finally in the reverse. After reading the next few pages you'll see for yourself why one-speed light means simultaneity is relative and why lengths necessarily contract with motion. When you're ready, close your eyes, take a deep breath, and turn the page.

Episode II

LIGHT GOES AT ONE SPEED
SO TIME HAS TO BE RELATIVE

AS DOES SIMULTANEITY,
WHICH MEANS LENGTHS MUST
CONTRACT WITH MOTION

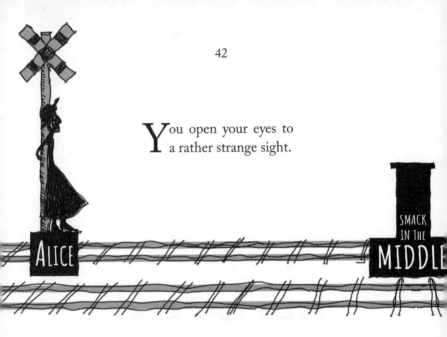

You open your eyes to a rather strange sight.

Before you are two parallel train tracks.
Sandwiched between them are two posts.

Tied to each post is an outlaw. The one to your left is Alice; the one to your right, Bob.

Smack in the middle is a narrow podium.

There's also something in your hand. A peculiar gun of some sort. Your gun has not one but two barrels, pointing in opposite directions. On the top there's a big flat button that says "FIRE."

You press it tentatively and…

<—<—<— **Bzzzp!** —>—>—>

…two laser-like light pulses explode from either side. Damn! You'll have to be careful!

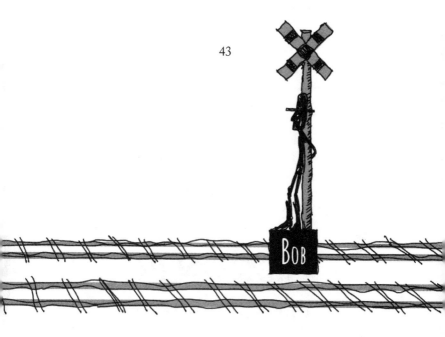

You scan the gun for instructions. And there, inscribed across the length of its twin barrels, you see this:

LIGHT GOES AT ONE SPEED

So, both light pulses go at the same speed, meaning they'll always be the same distance from the firing point. You'll keep that in mind.

You decide to investigate the podium. You make your way there, carefully stepping over the first set of tracks, to hoist yourself up on the small stage.

On the top of the podium there's a hole marked "Midway." It seems to be the right size for your gun handle. In fact, it's a perfect fit. With a CLICK the gun satisfyingly locks into place, its twin barrels aimed precisely at Alice and Bob on either side.

And then you see this note:

MISSION #1
Hit Alice & Bob Simultaneously

You look around. There's no one else in sight. It must be for you. And anyway, the setup is perfect. Your gun is properly aligned, positioned smack in the middle between them. To hit both simultaneously with one-speed light pulses all you need to do is press the "FIRE" button. Nothing could be easier.

Your index finger is nearly touching the gun's smooth flat button, when you feel something. A slight rumbling under your feet. You look along the tracks past the Alice outlaw to see what it could be. Indeed, there's a white flat-bed freight train coming towards you. Does it matter? It shouldn't interfere with the success of your mission. As you reach to press FIRE, you blink...

The White Train

…and open your eyes to see your hand pushing through empty space where the gun was a moment ago. You look around, slightly annoyed. The gun is gone. And you're no longer even on the podium. Somehow you've been transported to the front of the fast-moving white freight train. You deduce that you're now riding the very train you saw when you were about to fire the gun. Indeed, just ahead, you can clearly see the two outlaws on either side of the podium, Alice closer and Bob just beyond her, getting nearer to you every second. And there's something in your hand.

It's a note. But not the one from before. This one, Mission #2, is asking for something quite different.

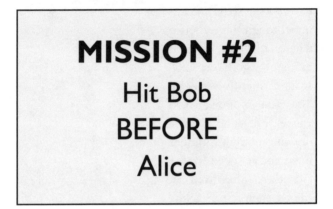

So, you have to hit the outlaw that's further away first and then the closer one. And from the moving train no less. Tricky. But where's your gun? Then you remember. You left it on the podium midway between Alice and Bob.

You can make them out more clearly now, Alice closing in, Bob not far behind, the gun on the podium, smack in the middle between them. One-speed light pulses fired from there will hit Alice and Bob simultaneously. Won't they?

ALICE

MI

S
I

You stick out your left thumb, imagining it's you. You hold the book up behind your thumb with your right hand and move it from right to left, in the same direction as the train tracks and ground are moving relative to you. You imagine the gun firing the instant thumb-you lines up with the podium. The emitted light pulses, oblivious to the fact that the gun that fired them is on the move, go at the same speed, traveling away from the firing point behind your thumb at the same pace, always exactly the same distance away from thumb-you. All the while, Bob advances towards the light pulse heading at him, even as Alice speeds away from hers. From your perspective on the white train, a double shot fired from the podium will hit Bob before Alice, neatly accomplishing Mission #2!

And that's when you notice it. Just beyond Bob on the other track coming towards you is another flat-bed freight train like the one you're riding, only black.

You blink and...

The Black Train

…open your eyes to find yourself riding that very same black freight train. This time you're heading towards the two outlaws from the opposite direction. Straight ahead is Bob, Alice not far behind. Just beyond them, on the other track, is the white train, coming at you quickly. The note in your hand, and its message, doesn't really surprise you. It reads…

MISSION #3
Hit Alice
BEFORE
Bob

So, the opposite order from the order in Mission #2, but like last time, you have to hit the further outlaw first. You play it all out but again, this time in the opposite direction. Yes, it can be done! The shot from the podium can accomplish all three missions!

You blink and…

The Shot from the Podium

…you're back on the podium, the gun exactly where you left it. And here comes the white train from one side, the black train from the other, in perfect symmetry, converging and passing each other right in front of you. You fire as the back ends of both trains' leading train cars pass by. The light pulses go in opposite directions from the firing point at the same speed.

Bzzzzzp!!

Alice and Bob, equally distant
from where the shot was fired…

are hit **simultaneously!** On contact both
vaporize, leaving their marks on the passing trains.
"A" for Alice and "B" for Bob.
Mission #1 accomplished!

HIT!   HIT!

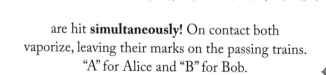

The Shot from the White Train

But how would that same shot play out from the white train? Even as the thought enters your mind you are transported there …

BZ

Both light pulses emitted from the firing point go the same speed in opposite directions, heedless of the fact that the gun that fired them is moving relative to your train. All the while, Alice recedes from the pulse that chases her.

←—<—<—<—<—<—

HIT!

And even though Alice is on the move, the light pulse, ever gaining, eventually catches up. She too is hit and leaves her mark. **The very same shot that hit Alice and Bob simultaneously also hit Bob before Alice…**

←—<—<—<—<—<—<—<—<—<—<—<—<—<—

A

51

… to the white train's leading car, at precisely the moment the shot is fired!

ZP!!

Bob, on the other hand, runs smack into his …

HIT!

… getting hit about three quarters of a train car from where the gun was fired, leaving his mark, B for Bob, on both trains.

… thereby accomplishing both **Mission #1** and **Mission #2**! And that's when you notice something a little odd.

So, you freeze frame to take a closer look.

No, you weren't imagining it. The distance between the A and B marks left by Alice and Bob when they were hit is shorter on the black train than on your white train. But why?

You play it out in your mind's eye. You think back to the moment when Bob was hit. At that point the two resulting B marks were perfectly aligned. Then some time passed during which the black train continued on its way, naturally taking its B along for the ride—until, as you see in the frozen moment before you, Alice's light pulse caught up to her, resulting in two A marks, one on the white train and one on the black, opposite each other for the moment, just as the B marks were at the instant Bob was hit.

Okay, so it does makes sense. In fact, it couldn't be any other way. And yet, somehow you can't quite get over the feeling it must mean something. But what?

HIT!

You look at the A and the B
on your white train marked
where Alice and Bob
were hit.

That's
the distance
between the two
events on the white train.

The distance between the same two
events is shorter on the black train, because
in the time between when Bob was hit and
when Alice was hit, the black train was moving.

And that makes you wonder: What if there was no
time between? From the podium that same shot hit
Alice and Bob simultaneously, meaning no time at
all passed between when Alice and Bob were hit, in
which case the distance between A and B is necessarily the same on both trains.

And that makes you wonder …

HIT!

HIT!

The Shot from the Black Train

... about the black train. Even as the thought
crosses your mind, you find yourself there, on
the black train's leading car, at the very moment
the shot is fired, both one-speed light pulses
emitted from the firing point.

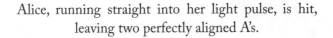

Alice, running straight into her light pulse, is hit,
leaving two perfectly aligned A's.

Eventually, the other light pulse catches up to the
receding Bob, and he too is hit, thereby completing
Mission #3, to **hit Alice before Bob**,
with the one shot.

And in the time between, the white train, constantly in motion, takes its A along for the ride, so that by the time Bob is hit, the distance between the white train's A and B is less than the very same distance on the black train you're riding.

But how does that make any sense? You decide to review the facts. One shot was fired midway between Alice and Bob. On the white train you saw that the light pulses from the shot hit Bob before Alice. On the black train you saw the very same light pulses hit Alice before Bob. And on the podium, you saw them hit Alice and Bob simultaneously. There's no way around it. Mission Impossible was in fact accomplished with one shot. All because light goes at one speed. But what if ...

What if Alice and Bob had met their ends for some other reason? What if they had instead, say, spontaneously combusted? What if their demise had nothing to do with one-speed light at all? Would Alice's time of death still precede Bob's from the passing black train? Would Bob's still precede Alice's from the passing white train?

You play out the scene from the podium, this time imagining the light fired from the midpoint just barely missing both outlaws, who, quite by chance, happen to evaporate at the very moment the pulses whiz by. Now you imagine the same scene, only this time you don't press FIRE. There are no light pulses, and the outlaws simultaneously combust all on their own.

It would make no difference. One-speed light doesn't change the facts. It merely reveals them. The relativity of simultaneity isn't a feature of one-speed light. It's a feature of time exposed by one-speed light. That's just how time works. The fact that light goes at one speed doesn't make time like that; rather, it affords a fleeting glimpse into time's otherwise-hidden nature.

And there's the other thing: the distance between Alice and Bob, captured in the gap between the A and B marks created when they were hit. You imagine that distance, about two train cars in length. Enough room for four elephants or so. On the white train, you saw the black train's A and B fit well within the white train's letters. Anything in that space would have to fit too.

And on the black train, you saw the white train's A and B fit well within the black train's letters. That means that on the white train that same distance on the black train is shorter, and on the black train that same distance on the white train is shorter. All because in the time between when Alice and Bob were hit the other train was in motion.

And if the trains were going faster? Faster-moving outlaws would see the approaching outlaw hit sooner and the receding one later, resulting in a greater distance between the A and B marks on the train you're riding and more time for the letters on the other train to get closer together. Faster trains mean more train cars have to fit into less space!

And if there had been no A or B? No light-struck outlaws at all, for that matter? Would it change anything? Would it change the fact that the moving black train is shorter measured against the white train and vice versa? It wouldn't. That just how space works. One-speed light doesn't make space like that. It just exposes its true character.

You imagine a distance between two points, say downtown New York and Tokyo, and everything in between: people, high rises, football stadiums, oceans, whales, mountains, and, yes, probably even elephants. If you zoom by fast enough, all that stuff would compress into the width of your little finger!

You take a deep breath, hold it for a moment, and slowly exhale. Space and time are not what they seem. They are not rigid and absolute. They are fluid, expanding and contracting with complete disregard for the unyielding character we attribute to them as the result of our everyday experience of the world. Light goes at one speed: these words are a window. A window with a view into the delightful secret life of space and time, should anyone care to look through it.

End scene.

It became clear that to speak
of the simultaneity of two events
had no meaning except in
relation to a given
coordinate system.

—Albert Einstein

L ight goes at one speed, so time has to be relative, as does simultaneity, which means lengths must contract with motion. This is as true in the real world as it is on imaginary trains because it's a direct consequence of the fact that light goes at one speed.

Einstein's insight that simultaneity is relative was the key to unlocking the apparent contradictions of one-speed light. Armed with a conceptual framework for resolving these seeming paradoxes, he would go on to uncover the astonishing relationship between distance, time, and relative velocity.

In the next episode you'll see exactly how relative velocity affects the duration of a second and the length of a train car, and you'll understand why one-speed light requires distances and times to come out as they do. You'll even get the chance to accelerate or slow down the trains for yourself to play out these effects at any speed.

All this will be seen by considering a light pulse fired straight up from the black train at an overhead balloon. You'll experience this same balloon-popping shot, first from the black train, then from the white train, and again, for the last time, from the white train, where you'll fire a diagonal shot that travels alongside the black train's vertical one, revealing the exact contraction factor of space and time.

So, what do you need for this ride? No high school math or calculators required. Two trains moving relative to each other at half the speed of light, some string, a few balloons, and an imaginary ruler will do the trick. When you're ready, close your eyes, take a deep breath, and turn the page.

Episode III

LIGHT GOES AT ONE SPEED
SO TIME HAS TO BE RELATIVE

AS DOES SIMULTANEITY,
WHICH MEANS LENGTHS MUST
CONTRACT WITH MOTION

BY A VERY PARTICULAR AMOUNT

PART I
TIME DILATION
Exactly How Much Time
Passes on Each Train?

You open your eyes to find yourself on the white train, a light gun in hand. There's no black train in sight. Everything is still and calm.

You look down.

At your feet are lengths of string. Lots and lots of them, each extending from one end of your train car to the other. And there's something else.

A big box of balloons.

You pick one up. It looks like an ordinary balloon. You press the balloon's opening to your lips.

Puff, puff, puff, and just like that it's fully inflated. You reach for a string, tie off the end, and let go. To your delight the balloon gracefully floats up and up.

You catch the end of the string just in time. The balloon, tied to a train-car-length string, now floats above your head precisely one train car length away.

You decide to inflate another. But what to do with this balloon? Luckily, right next to you, on the very edge of the white train, directly between two train cars, is a balloon stand with a clasp. You secure the first balloon and reach for a second.

Several hours pass by as you inflate and tie off one balloon after another. Eventually there are none left in the box.

You look up.

You were so absorbed in your task you hadn't noticed that the balloons arranged themselves, single file, into an arc that extends to the far ends of the two train cars on either side of you.

Wow.

After you admire your handiwork for a moment, the light gun catches your eye.

Could you hit a balloon?

You'd sure like to try!

You brace the gun on the balloon stand.

Which balloon should you pick? As you shift your gun's sights from one balloon to another, a thought pops into your head.

Since each balloon is attached by a train-car-length string, your shot will take exactly one second to hit any balloon on the arc.

With that in mind
you choose your target,
a striped balloon that floats
halfway over the train car to your right.

You aim your gun along the balloon's string …

and …

begin …

to …

squeeze …

the …

trigger …

The light pulse fires—Bzzzp!—into empty space?!?

Your perfect balloon arc is gone!

You look around in confusion. You're no longer on the white train at all. Somehow, you've been transported to the front of the black train. And there's the white train zooming by at its familiar dizzying speed.

While there's no balloon arc on the black train, there is a balloon stand between two train cars on the edge next to the white train. And at your feet, a single string exactly a train car long, and one uninflated balloon.

You blow up the balloon and secure it to the stand with the string. It now floats directly overhead exactly a train car length away.

Well, at least you still have a target to fire at.

Bracing your gun on the balloon stand,
you point it straight up and shoot!

Your aim is true! Your light pulse travels parallel to the string straight towards its target. In a quarter of a second it's a quarter of the way up, in half a second halfway, in three quarters of a second three quarters of the way, until—POP! The light pulse meets its mark precisely one second after being fired.

One-speed light makes the position of the pulse along the train-car-length string a perfect clock marking the passage of time between when the trigger is pulled and the balloon meets its end.

A thought pops into your head: What would this one-second shot have looked like from the moving white train?

And indeed you're about to find out. Even as the question forms you find yourself transported there, back to the white train, to the moment before the black-train you fired straight up at the overhead balloon. The single balloon is once more intact, and the black-train you is directly under it, light gun aimed at the target overhead, preparing to fire.

In passing, the black-train you pulls the trigger. The balloon-popping shot now plays out in dramatic slow motion as you look on from the white train.

You see the light pulse emerge from the gun and head directly towards its target. To your initial surprise the shot doesn't go straight up. Instead it travels parallel to the balloon's string, which, being attached to the black train, is moving along with it. Relative to the white train the light pulse moves, not vertically, as it does on the black train, but diagonally.

As the scene unfolds in slow motion, you play with the gun in your hand, maneuvering it to match the angle of the receding light pulse. You get it just right so that if you were to pull the trigger, your shot would follow the path of the light pulse fired from the black train at the black train's overhead balloon. Suddenly you get a déjà vu feeling.

There's something about the position of the gun that's awfully familiar. Why is that? Then it comes to you. The gun is at the same angle as when you held it while aiming at your chosen target balloon on the balloon arc. And as the light pulse fired straight up from the black train hits the black-train balloon—POP!

—you find yourself still on the white train, but now transported back to where you started, under the center of the balloon arc. As before, your gun rests on the balloon stand, aimed directly at that very same striped target balloon floating halfway above the next train car over, perfectly intact and ready to be popped. You note that your gun is indeed also at the exact angle to fire a shot that takes the same path as the one fired straight up from the moving black train. You're clearly here to witness the black-train balloon-popping shot one final time, but this time you'll see what happens when you fire at your target balloon on the white-train balloon arc at the same time as the black-train you fires straight up at the black-train balloon in passing. Those shots, fired together, will stay together, because light goes at one speed. This elaborate setup suggests that double shot must pack a serious punch!

You glance over your shoulder, and indeed, it appears you're about to find out, because the black train, speeding alongside your white train, is carrying the black-train you, who is getting closer every second and clearly preparing to fire at the once-more-intact overhead balloon tethered to the black train by a train-car-length string. You brace yourself.

In the instant you pass one another, the two guns so close they're almost touching, you both FIRE! The scene plays out in cinematic slow motion.

Two one-speed light pulses, the one you fired diagonally from the white train, and the other fired straight up on the black train by the other you, emerge in tandem and make their way together at the very same angle along your target balloon's string, neither able to overtake the other because light goes at one speed. In a quarter of a second the light pulses are a quarter of the way, in half a second halfway, in three quarters of a second three quarters of the way, until, one second after being fired—POP!—both light pulses make contact with the target balloon on your arc.

Freeze frame! The action arrests at the instant of the pop.

You survey the scene from the white train.

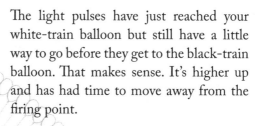

The light pulses have just reached your white-train balloon but still have a little way to go before they get to the black-train balloon. That makes sense. It's higher up and has had time to move away from the firing point.

The fact that the light pulses have gone the full distance of your balloon's train-car-length string tells you one second has passed on the white train. The black-train you, going at half the speed of light, has gone half that distance and is now exactly halfway across your train car directly under the popping balloon. But how much time has passed on the black train? Clearly less than a second. You can see that because the light pulses are only partway up the black-train balloon string. But exactly how far has the light gone up the black-train string in one white-train second? Because that's exactly how much time has passed on the black train.

If only you had some way to know. You reach into your pocket and smile. Of course. It's a measuring tape. You measure how far the light pulses have gotten towards their black-train target.

They are currently 86.6% of the way to the black-train balloon. That's a very specific number.

You review the facts.

On the white train a full second passed between when the guns were fired and the light pulses hit the target balloon on the arc. But not on the black train. The still-intact black-train balloon is evidence of that. So, less time passed on the black train between firing and popping. You look at the position of the light pulses along the black-train balloon string. To be precise, 86.6% of a black-train second passed between events separated by one white-train second.

Interesting.

You'd already seen one-speed light reveal time passing at different rates, but this is new. Now you're looking at exactly how much time has passed on the black train—86.6% of a second.

That number captures the relative rate of the passage of time between two events in a half-the-speed-of-light setup. But what if the trains were going at a different speed relative to each other? Either slower, or faster. How much time would pass in that case?

You imagine the same scenario speeded up, with the black train moving at three quarters the speed of light. At that speed, the black-train you would get three quarters of the way across the white train car in one second, which is also when a light pulse fired at the moment of passing would reach any balloon on the balloon arc.

Light pulses fired together stay together, so if the black-train you fired straight up from the midpoint of the arc, the light pulse would reach a balloon on the white train AND be somewhere along the black train's string after one white-train second—POP! You get out your tape measure. At three quarters the speed of light 66.1% of a black-train second passes between firing and popping—events separated by one white-train second.

POP!

Just for fun you try running the numbers at a few more speeds, making the trains go faster and faster. At 80% the speed of light .6 of a second passes on the black train. At 90% the speed of light, .435 of a second. What if the trains were moving at 99.99% the speed of light relative to each other? At that speed practically no time passes at all on the black train between firing and popping!

Who would think specific numerical facts about the universe could be hidden in five little words?

Light goes at one speed.

You take a deep breath, when quite unbidden, a nagging little thought pops into your head.

What would all this look like from the black train?

And before you can even exhale…

PART 2
LENGTH CONTRACTION
The White Train Is Shorter
Than the Black Train

You find yourself back on the black train, your light gun pointing straight up at the balloon overhead, the white train racing by a narrow gutter away.

You're now about to experience the very same double shot from Part 1 but this time from the black train's perspective.

And indeed, it's not long before you spot the balloon arc on the white train in the distance, approaching rapidly.

Before you know it, the moment arrives. The white-train you, aiming at the striped target balloon in the balloon arc, is right beside you. You both fire in unison!

In slow motion you watch the twin light pulses emerge and begin their tandem journey straight up the vertical string that tethers your balloon to the stand. With each passing moment the white train, constantly in motion, takes its balloon arc along with it, so the position where the two strings cross continuously moves up and up, keeping pace with the traveling light pulses. Eventually the white-train target balloon crosses your vertical string 86.6% of the way up—POP!

Freeze frame!

You survey the scene from the black train.

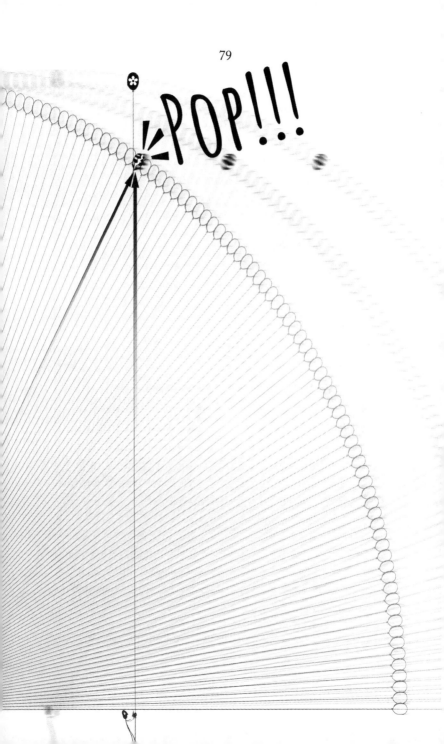

You glance down at your feet. Separated from your black train by a narrow gutter is the white train car's halfway point. The black train, moving at half the speed of light, has gone half the light's distance along the white train. That is where you find yourself now, directly under the popping white-train target balloon, half a white train car away from the white-train you.

But on the black train, less time has passed. The light has only gone 86.6% of the way up the train-car-length string, so only .866 of a second has gone by since the guns were fired. And the white-train you moving at half that speed has gone half that distance and so is now exactly 43.3% of the way along your black train car.

In this frozen moment the distance between the two yous necessarily measures half a white train car. In this frozen moment the distance between the two yous necessarily measures 43.3% of a black train car. Furthermore, this is a result of the fact that all train cars, black or white, are the very same length with regard to one-speed light. Namely the distance light travels in one second.

And for half a white train car to fit into 43.3% of a black train car, the white train cars need to be shorter. In fact, they need to be precisely 86.6% of a black train car long.

You glance up at the position of the light pulses.

Events separated by a full white-train second fit into 86.6% of a black-train second.

A full white train car fits into 86.6% of a black train car.

The moving white train's length and time are contracted by the very same factor!

PART 3
HOW CAN IT BE?
The Black Train Is Shorter
Than the White Train

As you stand on the black train, stuck in the frozen moment, .866 of a second after the two guns fired, halfway across the white train car with the white-train you .433 of a black train car away, you begin to have second thoughts.

Why is the white train the shorter one? Why not the other way around? Aren't the trains moving at half the speed of light relative to each other? Couldn't the balloon arc just as easily have been on the black train? What if both trains had a balloon arc? Surely the white train can't be both shorter and longer at the same time.

You decide to get to the bottom of it, to tweak the scene so that it's the same on both trains. But first you simplify things a bit. For example, you don't need the entire white-train balloon arc.

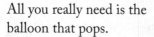

All you really need is the balloon that pops.

Now to make it symmetrical. Your black train gets a diagonal balloon on a train-car-length string floating exactly halfway across your black train car…

POP!!!

POP!!!

and the white train
gets a vertical balloon.

Now both trains have
two balloons each, one diagonal
and one vertical.

POP!!!

To keep it even, you imagine that both yous each fired two shots, one diagonal and one vertical, at the two balloons tethered to their respective stands.

From the black train, this moment, .866 of a black-train second after the shots were fired, reveals a scene in which the one-speed light pulses, having been fired together and stayed together, are all currently 86.6% of the way along both your black train's train-car-length strings.

The light pulse overhead is just popping the white train's diagonal balloon. Nothing new there. One white-train second has passed.

And the other strings? The light pulse, 86.6% of the way along the black train's diagonal string, is currently 75% of the way up the white train's vertical string. So three quarters of a white-train second has passed.

Wait.

Which is it? Has three quarters of a second passed on the white train or has one second passed?

Your heart beats a little faster.

The answer is both. You are seeing with your own eyes what Einstein saw.

The relativity of simultaneity.

You are seeing two moments, separated in time on the white train by a quarter of a second, occur simultaneously on the black train!

One moment, captured in the position of the light pulses on the white train's vertical string, occurs .75 of a second after the shots are fired. The other—POP!—after one white-train second.

And both moments, which happen a quarter of a second apart on the white train, occur at the exact same time on the black train, .866 of a second after the guns were fired, exposed right before your eyes by one-speed light!

Insane!

You're on the black train, standing under the white-train balloon frozen in mid-pop, when a wild thought comes into your head. What would happen if you stepped over onto the white train right here and now? What would the quadruple shot, .866 of a black train second after being fired, look like from the white train?

You decide to go for it! Crazy times call for extreme measures. And anyway, there's no one around to stop you! Leaving your black-train self behind, you gingerly step over to the white train. There, you did it. You're now standing half a white train car from the white-train you and directly across from the black-train you.

You look up.

The light pulses are exactly where they should be, 86.6% of the way up the black-train string and at the end of the white-train string, popping the attached balloon.

POP!!!

That makes sense.
After all, you've stepped
onto the white train one
second after the light
guns were fired.

POP!!!

(ALREADY POPPED)

And there's the white-train you. The light pulses fired along the vertical white-train string, having had a full second to make their way up its length, have reached the overhead balloon, which is also in mid-pop.

But what happened to the second black-train balloon? The one that hovered halfway across your black train car? It's already been popped. When did that happen? That's when it hits you. That balloon must have met its end .866 of a second after the shots were fired, at the moment it crossed the white train's vertical balloon string, just as the white-train balloon met its end on the black train. On the white train that moment occurred .134 of a second ago!

This white-train moment, one second after the light guns were fired, includes a sequence of events separated in time on the black train that here occur simultaneously, a sequence made all too tangible in featuring one popped and one intact black-train balloon!

Your mind reeling, you remember why you wanted to run the symmetrical double-pop experiment in the first place. On the black train the white train cars are shorter. But now you're on the white train. Is that still true? You make your way over to the frozen white-train you to check.

Standing directly under the popping white train's vertical balloon, you see clearly that one white-train second has passed. The black-train you, moving at half the speed of light, is now exactly halfway across the white train car.

However, more than a second has passed on the black train. After all, the black-train balloon popped .134 of a second ago. On the black train 1.154 seconds have gone by since the shots were fired, meaning the white-train you, moving at half the speed of light, has gone half that distance and is now .577 of the way across the black train car.

In other words, .577 of a black train car fits into half a white train car. Meaning the black train cars are 86.6% the length of the white ones. On the white train, the black train cars are shorter.

You stumble back over onto the black train.

Instantly, the white train with its shorter train cars continues at its usual dizzying pace, traveling at half the speed of light relative to your longer black train.

PART 4
ADDING VELOCITIES
Summing It All Up

You sit on the black train with the white train whipping by and think about what you just experienced.

You saw events separated in time on one train occur simultaneously on the other. You saw that from the white train the black train is shorter, and from the black train the white train is shorter, the length contracted by the same factor by which time is dilated.

All because light goes at one speed.

What other strange features of the universe could one-speed light still reveal?

And even as the question forms in your mind you see something moving in the distance on the other side of the speeding white train. It's a red train, going in the same direction as the white one, but even faster!!!

As you look on, the action mercifully begins to slow down, and down, and down, until eventually the scene unfolds at a crawl. You now watch it all play out from the black train in almost-relaxing slow motion.

At this reduced frame-rate the white train appears to inch along. You can now clearly see that its shortened train cars fit well within the lengths of yours. And there's a you standing on the white train at the end of a white train car, some distance away but "slowly" getting closer at half the speed of light.

Just on the other side of the white-train you is the front end of the even-faster red train moving in the same direction as the white train. And at the back end of its leading train car, a new red-train you! You look on with interest as the two yous get closer.

After a few moments it's quite clear that the red train on the opposite side of the white train is going faster than the white train, because while the white-train you is currently ahead, the red-train you is definitely catching up. In fact, it looks as if the red-train you will pass the white-train you at the exact moment they both reach the spot where you're now standing on the black train.

And indeed, that is exactly what happens.

The white-train you and the red-train you converge for an instant directly in front of you. As they pass, you notice that the red train cars are shorter still than the white ones, and by a significant margin. You suppose that makes sense since the red train is evidently going faster than the white.

But how much faster?

The action continues in slow motion until the white-train you, going at half the speed of light, is halfway across your train car. One second has passed. **Freeze frame!**

Perfect! You can now answer your own question. You'll be able to see how fast the red train is going by looking at how far the red-train you has come along your black train car in one second.

You walk
past the white-
train you frozen
at the halfway point.
It turns out that the red-
train you is exactly .8 of the
way across your black train car.

That means the red train is going at .8
the speed of light relative to your black train.

You take in the shortened red train cars. How long are they, exactly? You recall that you already went to the trouble of working out the time dilation factor for a train going at .8 the speed of light in Part 1 of this very thought experiment while standing under the balloon arc. At .8 the speed of light, .6 of a red-train second passes between events separated by one black-train second. Since lengths contract by the same factor by which time dilates, you know the red train cars are exactly .6 the length of yours. Which looks about right.

The red-train you going at 80% the speed of light has made it .8 of the way across your train car in one second. And how far along the white train car has the red-train you come in that time? Because that will tell how fast the red train is going relative to the white one. You look to see.

Evidently the red-train you is .433 of the way across the shortened white train car.

So the red-train you has gone .433 of a white train car in one second. Doesn't that mean the red train is going .433 the speed of light relative to the white train? Can that be right?

As you scratch your head, you realize what's bothering you. The red-train you has indeed gone .433 of a train car in one black-train second. In order to figure out how fast the red train is going relative to the white train, you need to know how far it's gotten in white-train time.

So how much time has passed on the white train since the red-train you went by? That's easy! You just saw it for yourself in Part 1 when you measured how far the light had gone along the black train's vertical balloon string when the white-train balloon popped. On the white train .866 of a second has passed. Meaning that the red-train you has gone .433 of the way across the white train in .866 of a white-train second; .433 is exactly half of .866.

The red train is, then, going at half the speed of light relative to the white train.

And the white train is going at half the speed of light relative to your black train.

And, remarkably, the red train is going at .8 the speed of light relative to the black train.

Which means that adding half the speed of light to half the speed of light doesn't get you the speed of light. Length contraction and time dilation take care of that. Half the speed of light plus half the speed of light results in .8 the speed of light.

And if there were another train, on the other side of the red one, going at half the speed of light relative to it, how fast would it be going relative to you?

As if made to order, such a train comes into the picture, a purple one, just beyond the red one. On the red train .6 of a second has passed. The purple-train you, going at half the speed of light relative to the red train, has made it .3 of the way across the red train car in that time. Which puts the purple-train you .929 of the way across your black train car. The purple train is therefore going at 92.9% the speed of light relative to your black train.

And now there's a new green train on the other side of the purple one, going at half the speed of light relative to it. Since only .37 of a second has passed on the purple train, the green-train you, going at half the speed of light, has gone half that distance, .185 of the way across its length, putting the green-train you .975 of the way along your black train car. Meaning that the green train is going at 97.5% the speed of light relative to your black train.

On every subsequent train less time has passed, and the train on its other side, going at half the speed of light relative to its neighbor, has only gone half that distance along its increasingly contracted length. No matter how many trains there are, each going an additional half the speed of light relative to the one next to it, no train will ever reach the speed of light relative to you. Time dilation and length contraction make sure of that.

Each extra train will get a little closer to going the speed of light but the difference is a smaller and smaller percentage of a shorter and shorter train car, so no matter how many trains there are, none will ever quite reach the speed of light!

Built right into the fact that light goes at one speed is another startling truth:

THE UNIVERSE HAS A SPEED LIMIT, WHICH IS THE SPEED OF LIGHT!

You inhale as you try to make sense of it all, but even before you can exhale, the action resumes in real time, with a now infinite number of trains zooming along in the same direction, each going at an additional half the speed of light relative to the one next to it, yet none ever quite reaching the speed of light relative to you.

End scene.

The most important upshot of the special theory of relativity concerned the inert masses of corporeal systems … and this led straight to the notion that inertial mass is simply latent energy.

—Albert Einstein

Light goes at one speed, so time has to be relative, as does simultaneity, which means lengths must contract with motion by a very particular amount, revealing the universe's speed limit, which is the speed of light!

And all of this leads us to the topic of the final episode, which, in Einstein's own words, is "the most important upshot of the special theory of relativity," namely, the exchange rate between energy and mass.

How did Einstein come to put two seemingly unlike things on opposite sides of the world's most famous equation? Why is a very large, very specific amount of energy equal to a very small, very particular quantity of mass, and what does the speed of light have to do with it?

The surprising relationship described by $E = mc^2$ shakes out when one-speed light is invited to the same party as a law of physics called the conservation of momentum.

An object's momentum is its mass times its velocity. Conservation of momentum means that in a collision, the total momentum of two objects before and after they crash remains the same.

Here's where it gets interesting. When the velocity part of momentum takes length contraction and time dilation into account, the law of conservation of momentum, which no one had really touched since Newton came up with it, becomes a veritable bombshell!

Are you ready to convert a huge amount of energy into a tiny bit of mass and experience the most important upshot of the special theory of relativity for yourself? Then take a deep breath, close your eyes, and turn the page.

Episode IV

LIGHT GOES AT ONE SPEED
SO TIME HAS TO BE RELATIVE

AS DOES SIMULTANEITY,
WHICH MEANS LENGTHS MUST
CONTRACT WITH MOTION

BY A VERY PARTICULAR AMOUNT

REVEALING THE EXCHANGE RATE
BETWEEN ENERGY AND MASS

$$(E = mc^2)$$

PART I

You open your eyes to find yourself on the white train. You're facing the black train, which is whipping by at its familiar dizzying speed. Something is moving behind you. You spin around.

There's another black train!

This time you're sandwiched between two black trains, each going at half the speed of light relative to your white train, but in opposite directions. You feel almost overwhelmed as you try to get your bearings.

Which is probably why it takes you a moment to realize there's something in your hand.

An egg? A quick glance reveals that it's not an egg but a plastic egg-shaped container with a Putty Time logo. You open it up. Inside is a small gray uninteresting-looking glob of putty.

You squish and stretch it. It appears to be ordinary putty, the kind you played with as a kid. And there's a slip of paper in the container.

You unfold it.

It says:

> # MOMENTUM IS CONSERVED!
>
> The total momentum of two objects before a collision is equal to the total momentum of the two objects after the collision.

You vaguely recall the law of conservation of momentum from somewhere. But what's it doing in a putty container? And what "two objects" does the message refer to? What "collision"? The thought makes you a bit nervous. Maybe there's more on the other side. You flip it over and see:

> Putty (mass 1) going one train car per second has 1 unit of momentum.
>
> # USE PUTTY TIME!

"Use putty time"? You look closely at the container, and indeed, right on top is a small one-second timer, complete with a tiny start/stop button.

Curious. And what about the rest of the message? "Putty (mass 1) going one train car per second has 1 unit of momentum." What can it mean?

Apparently, the putty in your hand has a mass of 1. Okay. But what about the bit that mentions one train car per second? That's the speed of light! Surely a glob of putty can't go that fast. Nothing but light can. But evidently that's how fast the putty has to be going to have one unit of momentum. Maybe one unit of momentum is just some unattainable limit like the speed of light? Anyway, the fact that the putty has a mass of 1 certainly makes things convenient. Momentum is mass times velocity. Any number times 1 is just that number. The putty has a mass of 1, so its momentum is just its velocity in the direction of motion.

You look at the putty in your hand. It isn't moving at all right now. It has zero velocity, so zero momentum.

You take in the two black trains. They go at half the speed of light. Doesn't that mean the putty in your hand is moving at half the speed of light relative to each of them? That's half the speed in the message. You suppose the putty, then, has half a unit of momentum on either black train.

You walk over to the very edge of your white train. A black train hurtles by a narrow gutter away. You balance the putty on your fingertips, lean in, and extend your arm. The putty is now perched a short distance above the floor of the fast-moving black train.

You look down. The train is a blur of motion, each train car passing by, one after another, so fast it's almost hypnotic. As your mind reels with the speed of it all, you recall that the moving black train cars are shorter than yours. Each is only .866 the length of a white train car, a fact you discovered shooting at balloons on the balloon arc in a previous episode. One-speed light means it could be no other way.

And that makes you wonder about the literal meaning of the message on the slip of paper:

**Putty (mass 1) going
one train car per second
has 1 unit of momentum.**

You just assumed the putty in your hand has half a unit of momentum on the black train because the black train is going half a train car per second relative to you. But the black train cars are shorter. Surely the length of the train cars affects the speed of the putty measured in train cars.

You imagine an extreme case with super-short black train cars, say one tenth the length of yours. In that case the putty in your hand would pass by not half a train car per second, as you'd initially supposed, but 5 train cars per second.

Of course, the black train cars aren't as short as all that, but still, at 86.6% the length of a white train car they are shorter, so the speed/momentum of your putty measured in black train cars must be proportionally more.

But how much more?

If only you had a timer. With a timer you could measure exactly how far the putty goes on the black train in one second.

Then you remember. There is a timer, right on the putty's egg-shaped container; the message specifically mentions it—use putty time! You position your thumb on the tiny stopwatch button and scan the length of the speeding black train for some kind of landmark. As if made to order, in the distance but approaching rapidly is a train car with a flag on either end. Perfect!

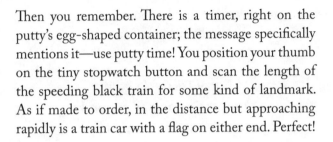

You brace yourself and wait. The marked train car is getting closer and closer. At the exact moment the flagged leading edge of the black train car lines up with yours—CLICK—you start the clock!

The action now proceeds in cinematic slow motion. As the timer counts down one second you see the front of the shortened black train car recede and its opposite end draw closer.

Tic-tic-tic-tic—DING!—precisely one second has passed—**Freeze frame!**

The leading flagged edge of the black train car, going half the speed of light, is now halfway across your train car. But since the black train cars are only .866 the length of yours, the putty balanced on your fingertips is well beyond the flagged black train car's halfway mark. In one second your putty has in fact crossed .577 of the black train car. Meaning that according to the message on the slip of paper, the putty has, not half, but **.577 units of momentum on the black train.**

You release the button on the putty-egg stopwatch. The action resumes once more in real time, with both black trains zooming by in opposite directions at half the speed of light.

PART 2

You cast your gaze along the length of one of the black trains. You spot something. Still far away but approaching rapidly is another you, getting closer and closer even as you look on. You can now see that the black-train you is leaning in with an arm extended far over your white train. And balanced on the tips of that you's two fingers is a small glob of putty.

You glance in the opposite direction. And there, equidistant in perfect symmetry with the first, riding the other black train, is yet another you, one arm reaching way over your white train car holding an identical putty lump also approaching at half the speed of light.

The two bits of putty are clearly on a collision course, destined to make contact at exactly the spot where your head is right now!

Weirdly the message from the putty instructions flashes before your eyes:

MOMENTUM IS CONSERVED!

The total momentum of two objects before a collision is equal to the total momentum of the two objects after the collision.

And just as the two lumps of putty are about to crash into you, you squeeze your eyes shut and drop!

THWUMP!!!

Then silence. Nothing. Are you dead?

You gingerly open one eye and then the other. Before you, perfectly stationary, is one largish lump of putty, evidently a combined mass made up of the two smaller lumps that were hurtling towards each other mere moments ago, now fused together as a result of their high-speed collision.

You think of the enormous amount of force each lump of putty had before they collided. Now they're perfectly still. All the energy of motion canceled out by an equal and opposite force at the instant of impact. You poke the lump. It seems harmless enough.

You pick it up.
It's heavier than you expected.

Were the bits of putty on the two black trains in fact the same size as yours? Were they the same size as one another, for that matter?

You think about it for a second. You know each glob of putty hitched a ride on a black train going at the same speed but in opposite directions. If one lump had been more massive, the bigger one would have had more momentum, meaning it would have taken the one with less momentum along for the ride. You imagine two unequal globs of putty going at the same speed, one the size of an elephant and the other the size of a mouse. On impact the elephant-sized one might have slowed down a bit, but it certainly wouldn't come to a stop. The resulting fused elephant-mouse putty lump would keep going in the same direction as the elephant-sized one. No, the fact that the momentum of the two globs of putty going in opposite directions at the same speed perfectly canceled out, bringing them to a dead halt, means they had the same mass before they collided.

A thought pops into your head: The momentum of the two lumps of putty, perfectly balanced from the perspective of the white train, wouldn't be symmetrical considered from the black train. On either black train one lump of putty would be stationary with zero momentum, with all the momentum coming from the lump of putty hitching a ride on the other black train. What would the very same collision have looked like from either black train? And before you can think it through you blink and ...

PART 3

Opening your eyes, you find yourself transported onto one of the black trains, which, now that you're here, feels perfectly stationary. You're facing the white train, which relative to you is zooming by at half the speed of light a narrow gutter away. And there, just on the opposite side of the white train, is the other black train, which, seen from your current perspective, is going in the same direction as the white one, only at a dramatically higher speed. Holy smokes! And you thought the white train was fast! The other black train absolutely puts it to shame. You look on in amazement. Since the white train is going at half the speed of light relative to you and the other black train is going at half the speed of light relative to the white one, it must be going at the SPEED OF LIGHT relative to you! Is that even possible?!? No, you remember. It isn't. You saw that for yourself under the balloon arc. You even specifically added half the speed of light to half the speed of light, which bizarrely but necessarily ended up being .8 the speed of light. Time dilation and length contraction, a direct consequence of one-speed light, make it so.

Regardless, the sight of a train going
at .8 the speed of light is impressive!

As you begin to get used to the sight you notice the plastic egg in your hand. Putty time. You open it up and take out the little lump of putty. You turn over the egg. Yes, there's the little timer. It's the same as the one you had on the white train.

Then you remember why you're here: to experience the putty collision from the black train's perspective. You look at the stationary putty in your hand; zero velocity, so zero momentum.

That means, from your position on the black train all the momentum of the putty crash that nearly killed you on the white train has to come from the moving lump of putty on the other black train, the one going at .8 the speed of light relative to yours.

You glance at the message on the little slip of paper that comes with the putty.

Putty (mass 1) going
one train car per second
has 1 unit of momentum.

According to the message, the putty going at .8 the speed of light would have .8 of a unit of momentum. Is that how much momentum your putty has on the other black train? But, you remember, it isn't. Because that wouldn't take the shortened cars on the insanely fast-moving black train into account. You think back to the balloon arc. You recall that the cars on a train going at .8 the speed of light contracted to a skimpy .6 the length of yours. Then how fast is the stationary glob of putty in your hand going in terms of the super-short train cars on the super-fast black train? Because that's how much momentum a glob of putty hitching a ride on that black train also has on yours.

You decide to find out.

Once again you position your thumb on the tiny stop-watch button, this time scanning the length of the speeding black train for some kind of landmark. There in the distance you spy a train car with a flag conveniently marking either end. You don't have long to wait, as it's coming towards you shockingly fast. At the instant the flagged leading edge of the black train car lines up with yours—CLICK—you start the clock! The action proceeds in dramatic slow motion.

As the timer counts down you see the front of the shortened black train car recede and its opposite end draw closer. Tic-tic-tic-tic—DING—one second has passed—Freeze frame! This time the leading flagged edge of the black train car going at .8 the speed of light is .8 of the way across your train car. But since the black train cars are much shorter, at only .6 the length of yours, the putty balanced on your fingers is now well beyond the opposite flagged end of the train car. In fact, it has passed the length of 1.333 black train cars in one second.

The momentum of the bit of putty in your hand measured in black train cars is therefore 1.333, a third more than the 1 unit of momentum the putty would have were it in fact going one train car per second! The putty's speed, when measured in contracted black train cars, is faster than light! You gasp and release the button.

The action resumes in dizzying real time.

PART 4

There's nothing for it. You know what you're here to do. You walk over to the edge of your black train and, balancing the putty on your fingertips, lean way over so that the lump is perched halfway above the speeding white train. You cast your gaze along its length, and indeed, as expected, there's the white-train you, wearing a look of abject terror, hurtling towards the stationary glob of putty on your fingertips.

And moving even faster on the other side of the white train, a short distance beyond the terrified white-train you, is the other black train with the other black-train you, the fingertip-perched lump of putty perfectly positioned to come crashing into yours!

It's all happening so fast! The white-train you, whose head is directly between the two putty lumps, is closer, but the black-train you is moving faster: clearly all three are on course to converge!

An instant before impact the white-train you ducks, both black-train yous jerk your hands back, and the bit of putty from the other black train, moving at .8 the speed of light, slams into yours! The two putty lumps combine into a single mass: the fast-moving one, with its 1.333 units of momentum, catching a ride on the other black train, has taken your previously stationary zero-momentum putty glob along with it for the ride.

Your heart racing, you watch the double lump, now going at the same velocity as the white train, recede at half the speed of light. You can just make out the white-train you gingerly poking the single fused putty mass, now stationary on the white train. Again, the message from the putty container pops into your head:

MOMENTUM IS CONSERVED!

The total momentum of two objects before a collision is equal to the total momentum of the two objects after the collision.

The total momentum of the two fused lumps of putty going at half the speed of light on the white train is, then, 1.333, just as it was before the collision.

But, hang on…

The momentum of one mass of putty going at half the speed of light is .577. You worked it out for yourself. The momentum of two such lumps is double that, so 1.154.

How can the double lump now riding the white train have 1.333 units of momentum?!? Where did the extra momentum come from?

Your pulse quickens.

Momentum is mass times velocity. The velocity is fixed. Each black train goes at half the speed of light relative to the white one. Period. To conserve momentum the mass of the double-putty lump has to go up. In order to have 1.333 units of momentum, the putty needs not two but 2.309 units of mass. That's a .309 difference. Each lump of putty somehow gained half that, namely .154 units of mass, on impact! Conservation of momentum means the crash that changed the putty's velocity also increased its mass by 15.4 percent!

PART 5

Where did that extra mass come from?!? You mentally replay the collision from the white train. Two identical putty masses zoom along at half the speed of light. They crash. On impact their velocity, and all their energy of motion, goes to zero. The change in velocity results in a 15.4 percent increase in the putty's mass.

The collision necessarily saw the loss of a very large, very specific amount of energy and a gain of a very particular, very small amount of mass. Energy lost, mass gained.

To be exact, the putty's energy of motion, going half the speed of light, was traded off for a 15.4% increase in its mass at the moment of impact.

So, 15.4% of the putty's mass is worth the energy that putty had going at half the speed of light. How much, then, is the whole piece of putty, all 100% of it, worth in terms of energy?

The putty's total mass is about six and a half times 15.4%. So the equivalent amount of energy in the entire lump of putty is six and a half times the energy traded off in the putty collision. That's enough energy to change the putty's velocity from half the speed of light to zero, about six and a half times over!

When you come to think about it, there's nothing special about the mass in putty per se. The six-and-a-half-times rule would apply to mass in any form. An egg is also made up of enough mass for its energy equivalent to change its velocity from half the speed of light to zero six and a half times. So is a grand piano or an elephant or a baseball.

It sounds like a lot of energy, but how much is it really? What would it actually take to stop, say, a baseball going at half the speed of light? The same amount of energy it would take to get it up to that speed in the first place, you suppose. But what would that energy amount to in real life? What else could that much energy do?

You decide to find out by doing a quick search.

Google: How much mass does a baseball have? Answer: Apparently about 140 grams. And how much kinetic energy does 140 grams have going at half the speed of light? You punch "140 grams" and "half the speed of light" into a kinetic energy calculator and get a really big number. Now to check it against something tangible. For example, the energy released by an atomic bomb, just for comparison's sake…

HOLY SMOKES!!!

Changing the velocity of a 140-gram baseball from 0 to half the speed of light, or vice versa, would take about thirty times the energy released by an atomic bomb!!!

And if it takes the energy equivalent of about 30 atomic bombs to increase the mass of a baseball by 15.4%, the total mass of a baseball, at six and a half times that, is then equivalent to about 200 atomic bombs worth of energy!!! The conversion rate of energy to mass is astronomically high!!!

You look at the putty with renewed respect.

Amazing.

You think about what it must have been like to be the first and, for a time, the only person to glean this intimate relationship between energy and mass. And how truly remarkable it is that these detailed specific facts about the world come into focus by unpacking the meaning of five simple words:

Light goes at one speed

You sigh.

Your mind is full but calm.

You gaze along the length of your train.

Far in the distance you can just make out a silhouette. It's a person tossing a baseball, up and down, up and down, with one hand, as if inviting you to play.

Is it another you?

No, somehow you're certain it isn't.

You know who it is.

He's been here with you all along.

You walk towards him.

You can now make out his wild, unkempt, almost-comical hair. You smile. His mischievous grin, unmistakable, is utterly infectious.

You easily catch the ball he tosses to you.

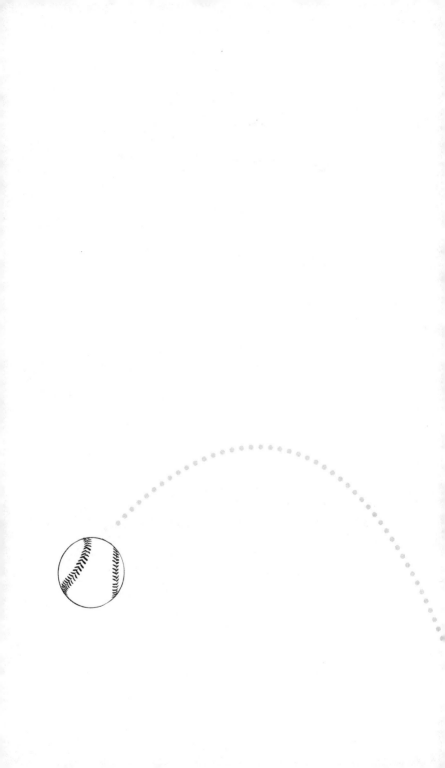

The final results appear almost simple … the years of anxious searching in the dark, with their intense longing, their alternations of confidence and exhaustion and the final emergence into the light—only those who have experienced it can understand that.

—Albert Einstein

Bonus Feature

THE CHICKEN OR THE EGG?
A TAKE ON EINSTEIN'S
TWIN PARADOX

In this bonus feature we will attempt to answer the age-old question Which came first, the chicken or the egg? and hopefully shed some light on Einstein's twin paradox in the bargain.

The twin paradox is a thought experiment in which twins are separated and one travels away from the other at a significant percentage of the speed of light. It usually goes something like this. Since time passes more slowly for the moving twin, she ages more slowly than the stationary one. But what if the story were told from the other twin's perspective so that the stationary one is moving instead? Doesn't the younger twin become the older one? Which twin is actually older? You'll soon see that in a symmetrical setup, it depends. It depends on which twin's moment you are in and when/where you are when you ask the question. If this last sentence gives you a relativity-of-simultaneity déjà vu, you're on the right track!

Our symmetrical setup will feature not two but four "twins" in the form of a batch of specially designed fast eggs that hatch exactly one second after their egg box is opened and mature into full-grown chickens .15 of a second after that.

There are two fast black eggs on the black train in the care of the black-train you and two fast white eggs on the white train in the care of the white-train you. To make it interesting both yous agree to freeze frame and send a newborn chick over to the other train the instant the eggs hatch.

You may find the numbers in this example awfully familiar. That's because in previous episodes you experienced similar events, seeing for yourself why the values must come out as they do. Feel like a challenge? Bring a pencil and paper along for the ride and use the balloon arc from Episode III to figure out the times and locations described in the scenario!

Are you ready to play out the chicken-and-egg twin paradox? Are you ready for your final train ride? Take a deep breath, close your eyes, and turn the page.

You open your eyes, to find yourself riding the white train, the black train zooming by next to you at half the speed of light. And you're holding something.

It's a box containing two fast white eggs.

Box in hand, you lean into the narrow gutter sepa rating your white train from the black one. Suddenly you see the black-train you approaching rapidly. That you is leaning over and clutching a box of fast black eggs. Before you know it, you pass each other, and the two egg boxes bump, knocking off both lids. The one-second hatch countdown is on!

The scene plays out in cinematic slow motion. The eggs on your white train mature even as the black-train you recedes with the other batch. After one second your eggs hatch and out pop two adorable fluffy white chicks—**Freeze frame!**

The black-train you, going at half the speed of light, is now exactly halfway across your train car. You know from your previous train episodes that .866 of a second has passed for the black-train eggs, so they have not yet hatched. In this white-train moment there are two newly hatched white chicks on the white train and two yet-to-hatch black eggs on the black train. As agreed, you place one of your white chicks on the black train. Since the black train cars are shorter, only 86.6% the length of yours, your chick ends up .577 of a train car away from the black-train you. From your perspective on the white train, the black train now carries one white chick and two black eggs.

But hang on. Your little chick is now standing .577 of a train car away from the black-train you. The white train was going at half the speed of light relative to the black train. That's half a train car per second. You and your chick are more than halfway along the black train. So, by the time you make it to this point, more than one second has passed on the black train. To be precise, going at half the speed of light the white train only gets this far after 1.154 black-train seconds. Meaning that by the time your white chick arrives on the black train, the black-train chicks will not only have hatched, they will already have matured into chickens! In this frozen black-train moment the new white chick arrives to join one fully grown black chicken on the black train. And what happened to the other black chicken? It's no longer on the black train at all! As agreed, the black-train you passed it over to the white train when it hatched .154 of a second ago!

You bid farewell to your white chick, and the action resumes in slow motion. For a time, your train's only passenger is the remaining lonely white chick, with the other three fast fowl catching a ride on the black train. Then, exactly 1.154 seconds after your egg box opened, the black-train you, now .577 of the way across your train car, freeze-frames the thought experiment to deliver an adorable freshly hatched black chick to join your remaining white now-chicken .154 of a second old on the white train.

With the black-train chick safely delivered, the action resumes in dizzying real time.

Despite feeling somewhat scrambled, you attempt to review what just happened.

On the white train…

You had two white eggs. After one second your eggs hatched. As agreed, you passed a newborn white chick over to the black train. For a time, there was only one white chick on your train. It matured into a chicken, after which point a newborn black chick was delivered. Your white train ended up with an older white chicken and a younger black chick.

And on the black train…

You had two black eggs. After one second your eggs hatched. As agreed, you passed a newborn black chick over to the white train. For a time, there was only one black chick on your train. It matured into a chicken, after which point a newborn white chick was delivered. Your black train ended up with an older black chicken and a younger white chick.

Even though your brain is now thoroughly poached, you can't help but wonder at the time warp you just experienced. Space and time are far more interesting than your day-to-day life had led you to believe! And just as you begin to come to terms with it all, a fact, true on both trains, pops into your head: for a time, you carried a single poultry passenger, yet you never carried three.

There are four fowl in this story.
Where did the missing chicken go?

Your eyes snap open.

End scene.

So which came first? Was it the chick or the egg, the chicken or the chick? Which is older, which is younger? The answer, astoundingly, is both, all, and neither because it depends on when/where you are when you ask the question! And where did the missing chicken go, anyway? The only place it could be. The missing chicken is on the other train!

This amazing but true story, brought to your mind's eye courtesy of one-speed light, is made possible by the astonishing nature of space and time, which delightfully stretch and shrink like a fantastical cosmic Slinky in a funfair house of mirrors!

Geek Notes

**FOR THE
SKEPTICAL RELATIVIST**

Episode I, Pages 13–39

The black train and the white train are two reference frames, moving relative to each other in one dimension along an infinitely long track.

In Newtonian mechanics, if the clocks in the two frames are synchronized at some zero time, the times in the two frames are the same: $t_b = t_w$. The relation between distance in the black train frame, x_b, and distance in the white train frame, x_w, after a time t is given by the expression

$$x_b = x_w - vt$$

In the special theory of relativity the relation between distances in the two reference frames, and between times in the two reference frames, is given by the Lorentz transformation:

$$x_b = \frac{x_w - \frac{v}{c} t_w}{\sqrt{1 - v^2/c^2}}$$

$$t_b = \frac{t_w - \frac{v}{c} x_w}{\sqrt{1 - v^2/c^2}}$$

where v is the velocity of the black train in the white train frame and c is the velocity of light in a vacuum, about 186,000 miles per second or 300,000 kilometers per second (light moves slower in air, glass, or water). The value of c depends on the units used, but only the ratio v/c is relevant. Our unit of length is the length of a train car. Our unit of time is about 100 nanoseconds or 10^{-9} seconds, since that's roughly the time it would take light to travel the length of a train car. Since in our units light moves one unit of length in one unit of time, $c = 1$, and the Lorentz transformation becomes

$$x_b = \frac{x_w - vt_w}{\sqrt{1 - v^2}}$$

$$t_b = \frac{t_w - vx_w}{\sqrt{1 - v^2}}$$

For convenience, we talk about light moving the length of a train car in 1 "second." Think of our "second" as shorthand for the amount of time it takes light to travel the length of a train car, about 100 nanoseconds.

In the white train frame the black train is moving to the right with velocity $v_w = 1/2$, and in the black train frame the white train is moving to the left with velocity $v_b = -1/2$. If we designate the beginning of a train car on each train as the zero position in the reference frame and synchronize the clocks on the trains when these zero positions coincide as the trains pass each other, then according to Newtonian mechanics, after 1 second the zero point in the white train frame (which is moving to the left in the black train frame) will be at the position

$$x_b = 0 - \frac{1}{2} = -\frac{1}{2}$$

in the black train frame.

The denominator $\sqrt{1 - v^2}$ in the Lorentz transformation is the length contraction factor, or time dilation factor. With $v = 1/2$, this is $\sqrt{3}/2$, which is approximately .866. To go from the Newtonian relation for the trains' positions in the two frames, $x_b = x_w - vt_w = x_w - \frac{1}{2} t_w$, to the relativistic relation, you simply divide by the contraction factor.

What amounts to the Lorentz transformation is derived in the balloon arc sequence in Episode III. In the preliminary sequence on p. 28, after the light has moved one train-car length in the white train frame, $x_w = 1$ and $t_w = 1$, so

$$x_b = \frac{x_w - 1/2\,t_w}{\sqrt{3}/2} = \frac{1 - 1/2}{\sqrt{3}/2} = \frac{1}{\sqrt{3}} \approx .577$$

which is (p. 30) "about halfway along the length of the black train car," and similarly, $t_b = \frac{1}{\sqrt{3}} \approx .577$, which is "about half a second" after the two light guns were fired.

Pages 30–31:

> One second has passed on the white train between when the guns were fired and the light pulses reached the end of the white train car—and as you can clearly see by the position of the light pulses on the black train, those events occurred in about half a black-train second.

At $t_w = 1$, the light pulse has moved one train-car length, so $x_w = 1$, from which it follows that

$$t_b = \frac{t_w - 1/2\,x_w}{\sqrt{3}/2} = \frac{1}{\sqrt{3}} \approx .577$$

which is "about half a black-train second."

Pages 31–32:

> In the one second it took the light to travel the length of the white train car, about half a second passed on the black train. But isn't the black train/white train setup symmetrical? Why then should less time pass on the black train than on the white train? What if the light guns had instead been fired together in the opposite direction?

> You imagine it: this time the light is shooting away from the rapidly receding black-train you. In this case, the pulses would pass over more than one black train car by the time they reach the end of the white one. So more than a second would have passed on the black train between events

separated by one white-train second. How can more *and* less than a second possibly pass on the black train in the one second it takes light to travel the length of a white train car?!?

The question is answered in Episode II, but the following remarks might help. When both the black train and the light are moving to the right relative to the white train, and one second has passed on the white train, the light is at the .577 mark on the black train, which means that .577 seconds has passed in black-train time ("about half a second"). Non-relativistically, an observer on the white train (the "stationary" train) would expect the light to be at the .5 mark after one second has passed on the white train.

To get to the .577 mark should take the light 1.154 seconds in white-train time (because after 1.154 seconds, the black train, moving at half the speed of light, moves a distance of $1.154 \times .5 = .577$ white-train car lengths in the direction of the light). Relative to white-train time, then, a black-train second must be shorter than a white-train second, in fact $\sqrt{3}/2 \approx .866$ of a white-train second: $1/.866 = 2/\sqrt{3} \approx 1.154$, which is to say that time is slowed down on the black train relative to the white train (or that "moving clocks run slow").

The same conclusion follows if we consider the black train moving to the right and the light moving to the left relative to the white train. After one second has passed on the white train, the light is at the 1.73 mark on the black train—more precisely -1.73, in the opposite direction to which the black train is moving— which means that 1.73 seconds has passed in black-train time ("more than a second").

If this is puzzling, note that—as explained in Episode II—events that are simultaneous for the white train are not simultaneous for the black train. So while the light reaching the .577 mark on the black train and the light reaching the −1.73 mark on the black train are simultaneous events for the white train (assuming the two light pulses are fired simultaneously on the white train by the two-sided gun in Episode II), they do not occur at the same time on the black train: some time passes between these events on the black train.

Non-relativistically, an observer on the white train (the "stationary" train) would expect the light to be at the −1.5 mark after 1 second, because the black train is moving away from the light at half the speed of light. To get to the −1.73 mark, which is further along the white train, should take the light 1.154 seconds in white-train time (because after 1.154 seconds, the black train, moving at half the speed of light, moves a distance 1.154 x .5 = .577 opposite to the direction the light is traveling while the light moves a distance 1.154 away from the starting point, giving a total distance between the starting point and the light as .577 + 1.154 = 1.73). So, again, relative to white-train time, a black-train second must be $\sqrt{3}/2 \approx .866$ of a white-train second.

Page 33:

> ... until at last it came to me that time was suspect!

The quotation is from Shankland, 1963, p. 48.

Page 36:

> Light is always propagated in empty space with a definite velocity c which is independent of the state of motion of the emitting body.

The quotation is from Einstein, 1923, p. 38.

Episode II, Pages 41–60

The motion of the two trains is symmetrical, both with respect to each other and with respect to the podium frame (the frame in which Alice and Bob are stationary). Suppose the velocity of the white train in the podium frame is v and the velocity of the black train in the podium frame is $-v$. To work out v, we need to add velocities relativistically, which is explained later in the book on pp. 90–97. If the velocity of the black train in the podium frame is $-v$, and the velocity of the white train in the black train frame is $1/2$, the velocity of the white train in the podium frame is

$$\frac{\frac{1}{2} - v}{1 - \frac{1}{2}v}$$

which we have taken as v. So, equating v with the preceding expression, we get

$$v = 2 - \sqrt{3} \approx .2679$$

We are now able to calculate the position of the marks A and B in the white train frame. Bob meets his end after a time $t_p = 1$ at position $x_p = 1$ in the podium frame. In the white train frame this position, the position of the mark B, is

$$x_b = \frac{1 - v}{\sqrt{1 - v^2}} = \frac{\sqrt{3} - 1}{\sqrt{4\sqrt{3} - 6}} \approx .7598$$

which is (p. 51) "about three quarters of a train car from where the gun was fired."

Similarly, we can find position A, the mark on the white train where Alice was killed:

$$x_a = \frac{-1 - v}{\sqrt{1 - v^2}} = \frac{\sqrt{3} - 3}{\sqrt{4\sqrt{3} - 6}} \approx -1.3161$$

The distance between A and B in the white train frame is $2/(4\sqrt{3}-6) \approx 2.0759$. By symmetry, this is also the distance between A and B in the black train frame.

In the white train frame the distance between the marks A and B on the black train is shorter than this because the mark B on the black train, originally lined up with the mark B on the white train, moves to the left during the time between the B event on the white train and the A event on the white train, as explained on pp. 52–53. To calculate the distance moved, we need to calculate the time, in the white train frame, between the two events: Bob's demise, yielding the B mark, and later, Alice's death, yielding the A mark. The time of the B event in the white train frame is the same as the distance of the B mark from the origin in this frame, which is .7598, because the velocity of light is 1. Similarly, the time of the A event is the same as the distance of the A mark from the origin, which is 1.3161. Subtracting these two times yields the time between the A and B events in the white train frame as $(4 - 2\sqrt{3})/\sqrt{4\sqrt{3} - 6} = .5562$; i.e., the A event occurs .5562 seconds after the B event in the white train frame.

During this time the black train moves relative to the white train with velocity $v = 1/2$, so in .5562 seconds the black train moves a distance $1/2 \times .5562 \approx$ .2781 relative to the white train, carrying the mark B on the black train (which originally lined up with the B on the white train at the position .7598 in the white train frame) to the position $.7598 - .2781 = .4817$ in the white train frame. This gives the distance, in the white train frame, between the A mark on the black train and the B mark on the black train as $1.3161 + .4817 \approx 1.7978$.

We could have calculated this from the distance 2.0759 between the A and B events in the white train frame using the contraction factor $\sqrt{3}/2 \approx .866$, which gives the same contracted distance: $2.0759 \times \sqrt{3}/2 = 1.7978$.

Page 59:

> It became clear that to speak of the simultaneity of two events had no meaning except in relation to a given coordinate system.

The quotation is from Einstein, 1954b, p. 230.

Episode III, Pages 61–99

Page 72:

> On the white train a full second passed between when the guns were fired and the light pulses hit the target balloon on the arc. But not on the black train. The still-intact black-train balloon is evidence of that. So, less time passed on the black train between firing and popping. You look at the position of the light pulses along the black-train balloon string. To be precise, 86.6% of a black-train second passed between events separated by one white-train second.

The length of the diagonal light path, the distance the black train has traveled until the balloon pops, and the length of the vertical string to the balloon form a right-angle triangle. From Pythagoras's theorem, we know that the length of the vertical is the square root of the square of the length of the hypotenuse (the diagonal), here 1, minus the square of the length of the base, here 1/2: $\sqrt{1 - (1/2)^2} = \sqrt{3}/2 \approx .866$. Similarly, time slows down in the black train frame relative to the white train frame. Specifically, after 1 second has passed on the white train, only .866 of a second has passed on the black train. Or, put differently, after 1 second on the black train, more than 1 second has passed on the white train, specifically $2/\sqrt{3} = 1.1547$ seconds.

Page 73:

> At three quarters the speed of light 66.1% of a black-train second passes between firing and popping—events separated by one white-train second.

Time slows down on the black train relative to the white train. At 3/4 light speed, a black-train second is $\sqrt{1 - (3/4)^2} \approx .6614$ of a white-train second.

Page 74:

> At 80% the speed of light .6 of a second passes on the black train. At 90% the speed of light, .435 of a second. What if the trains were moving at 99.99% the speed of light relative to each other?

At $v = .8$, $\sqrt{1 - (.9)^2} = .6$ of a second passes on the black train in 1 second on the white train. At $v = .9$, $\sqrt{1 - (.9)^2} \approx .4358$ of a second passes on the black train in 1 second on the white train. At $-.9999$, $\sqrt{1 - (.9999)^2} \approx .0141$ of a second passes on the black train in 1 second on the white train.

Page 85:

> The light pulse, 86.6% of the way along the black train's diagonal string, is currently 75% of the way up the white train's vertical string. So three quarters of a white-train second has passed.

Here we have a triangle with the length of the base $1/2 \times \sqrt{3}/2 = \sqrt{3}/4 \approx .433$ and the length of the diagonal (the distance the light has moved on the black train's diagonal string) equal to $\sqrt{3}/2 \approx .866$. It follows that the vertical length is $\sqrt{3/4 - 3/16} = 3/4$.

Page 86:

> You are seeing two moments, separated in time on the white train by a quarter of a second, occur simultaneously on the black train!

The two moments refer to events at different positions in the white train frame, half a train-car length apart. If this difference is denoted by Δx_w and the time difference by Δt_w, the time difference in the black train frame between these events is

$$\Delta t_b = \frac{\Delta t_w - v\Delta x_w}{\sqrt{1 - v^2}}$$

With $v = 1/2$, $\Delta t_w = 1/4$, and $\Delta x_w = 1/2$, it follows that $\Delta t_b = 0$.

Page 89:

> However, more than a second has passed on the black train. After all, the black-train balloon popped .134 of a second ago.

That's the difference in time between .866 of a second and 1 second.

Page 89:

> On the black train 1.154 seconds have gone by since the shots were fired, meaning the white-train you, moving at half the speed of light, has gone half that distance and is now .577 of the way across the black train car.

One second in the white train frame is $2/\sqrt{3} \approx 1.154$ seconds in the black train frame. In the white train frame, the black train is moving at $v = 1/2$, so the black train car length is contracted to $\sqrt{3}/2 \approx .866$ of a white train car length. So $2/\sqrt{3} \approx 1.1547$ contracted black train cars fit into the length of one white train car. Half of that is $1/\sqrt{3} \approx .5773$.

Page 94:

> At .8 the speed of light, .6 of a red-train second passes between events separated by one black-train second. Since lengths contract by the same factor by which time dilates, you know the red train cars are exactly .6 the length of yours.

If the velocity of the red train relative to the white train is $v = 1/2$, and the velocity of the white train relative to the black train is $v = 1/2$, adding the velocities relativistically gives the velocity of the red train relative to the black train as

$$\frac{\frac{1}{2} + \frac{1}{2}}{1 + \left(\frac{1}{2} \times \frac{1}{2}\right)} = \frac{4}{5} = .8$$

At a velocity $v = .8$, the time dilation factor, or length contraction factor, is $\sqrt{1 - v^2} = .6$. So time is slowed down on the red train relative to the black train. In one black-train second only .6 of a red-train second has passed. Similarly, red train cars are contracted to .6 of a black train car length.

Pages 95–96:

> On the red train .6 of a second has passed. The purple-train you, going at half the speed of light relative to the red train, has made it .3 of the way across the red train car

in that time. Which puts the purple-train
you .929 of the way across your black train
car. The purple train is therefore going at
92.9% the speed of light relative to your
black train.

The velocity of the purple train relative to the black
train is

$$\frac{\frac{4}{5} + \frac{1}{2}}{1 + (\frac{4}{5} \times \frac{1}{2})} = \frac{13}{14} \approx .929$$

Page 96:

Since only .37 of a second has passed on the
purple train, the green-train you, going at
half the speed of light, has gone half that
distance, .185 of the way across its length,
putting the green-train you .975 of the way
along your black train car. Meaning that
the green train is going at 97.5% the speed
of light relative to your black train.

In one second on the black train, only $\sqrt{1 - (\frac{13}{14})^2} \approx$
.37 has passed on the purple train. The velocity of the
purple train relative to the black train is

$$\frac{\frac{13}{14} + \frac{1}{2}}{1 + (\frac{13}{14} \times \frac{1}{2})} = \frac{40}{41} \approx .9756$$

Page 98:

The most important upshot of the special
theory of relativity concerned the inert masses
of corporeal systems . . . and this led straight
to the notion that inertial mass is simply
latent energy.

The quotation is from Einstein, 1954b, p. 230.

Episode IV, pp. 101–131

Page 104:

> Putty (mass 1) going one train car per second has 1 unit of momentum. **USE PUTTY TIME.**

In classical mechanics, the momentum of a mass m moving with velocity v is defined as mv. Relativistic momentum, in units where the velocity of light, c, is equal to 1, is defined as

$$\frac{mv}{\sqrt{1 - v^2}}$$

Dividing by the contraction factor $\sqrt{1 - v^2}$ amounts to using the classical definition and measuring time with a clock that moves with the mass, i.e., in this case, "putty time."

If the velocity of the putty relative to a train is $1/\sqrt{2} \approx .7071$, the time dilation factor is also $1/\sqrt{2}$:

$$\sqrt{1 - \left(\frac{1}{\sqrt{2}}\right)^2} = \frac{1}{\sqrt{2}}$$

So in 1 second in the train reference frame, only about $1/\sqrt{2}$ second has passed according to putty time. In 1 second of putty time, $\sqrt{2} \approx 1.4142$ seconds has gone by on the train. Since the putty has velocity $1/\sqrt{2}$ relative to the train, the putty moves $(1/\sqrt{2}) \times \sqrt{2} = 1$ train car in $\sqrt{2}$ seconds. That's to say, in 1 second of *putty time*, the putty moves 1 train car length, which means that the velocity of the putty using putty time is 1. Since the mass is 1, the momentum is 1, using putty time.

In other words, the message on the putty amounts to saying: "Putty (mass 1) going $1/\sqrt{2} \approx .7071$ of a train car per second has one unit of momentum *in train time*."

Page 110:

> Meaning that according to the message on
> the slip of paper, the putty has, not half,
> but **.577 units of momentum on the
> black train.**

The putty is stationary on the white train, so it moves
with velocity $v = 1/2$ relative to the black train. But
that's according to the time in the black train reference
frame. In the moving reference frame of the putty, time
is slowed down by the time dilation factor $\sqrt{1 - v^2} =$
$\sqrt{3}/2 \approx .866$. After 1 second on the black train only
.866 of a second has passed in putty time, so after
1 second in putty time $2/\sqrt{3} \approx 1.1547$ seconds have
passed in the black train reference frame. The putty
is moving with velocity $v = 1/2$ relative to the black
train; i.e., in 1 second of black-train time it moves half
a train car. In 1 second of putty time, the putty moves
a distance of $1/2 \times 2/\sqrt{3}$; i.e., $1/\sqrt{3} \approx .577$. So, using
putty time, the putty has a velocity of .577 in the
black train reference frame, and since the mass is 1,
the relativistic momentum is .577.

Alternatively, using the relativistic expression for
momentum, a mass of 1 moving with velocity $1/2$ has
a relativistic momentum of

$$\frac{\frac{1}{2}}{\sqrt{1 - (\frac{1}{2})^2}} = \frac{1}{\sqrt{3}} \approx .577$$

Page 120:

> The momentum of the bit of putty in your
> hand measured in black train cars is there-
> fore 1.333, a third more than the 1 unit of
> momentum the putty would have were it
> in fact going one train car per second!

A mass of 1 moving with velocity $4/5 = .8$ has momentum

$$\frac{\frac{4}{5}}{\sqrt{1 - (\frac{4}{5})^2}} = 4/3 \approx 1.333$$

Two lumps of putty, each with a mass of 1 gram, move towards each other at half the speed of light relative to the white train. When they collide, they stick together into a double lump and end up stationary with respect to the white train.

The momenta of the two lumps before the collision are equal and opposite, so they cancel out to zero. The momentum of the double lump is also zero, because it is stationary on the white train. Momentum is conserved: the total momentum before the collision is equal to the total momentum after the collision.

Now consider what this looks like in the black train reference frame. Before the collision, one of the lumps of putty is stationary on the black train, so its momentum is zero. The momentum of the other lump of putty going at $4/5$ the velocity of light is, as calculated above, $4/3$. So the total momentum before the collision in the black train reference frame is $4/3$. After the collision, the double lump, which is stationary on the white train, is moving with velocity $1/2$ in the black train reference frame. (A calculation, adding the velocities relativistically, gives $(0 + 1/2)/(1 + (0 \times 1/2)) = 1/2$.) To calculate the relativistic momentum of the double lump, we multiply the mass of the double lump —call it M—by the velocity, using a clock moving with the double lump. This velocity is

$$\frac{\frac{1}{2}}{\sqrt{1 - (\frac{4}{5})^2}} = \frac{1}{\sqrt{3}}$$

For momentum to be conserved in the black train reference frame as in the white train frame, we must have $M/\sqrt{3} = 4/3$; i.e., $M = 4/\sqrt{3}$.

Page 124:

> To conserve momentum the mass of the double-putty lump has to go up. In order to have 1.333 units of momentum, the putty needs not two but 2.309 units of mass. That's a .309 difference.

The original total mass was 2. So the increase in the total mass is $4/\sqrt{3} - 2 \approx .309$.

Relativistic energy for a mass m moving with velocity v is defined as

$$\frac{mc^2}{\sqrt{1 - \frac{v^2}{c^2}}}$$

In units in which $c = 1$, this is

$$\frac{m}{\sqrt{1 - v^2}}$$

This includes the relativistic kinetic energy, the energy of motion, and the energy associated with the mass, which is just m in units in which $c = 1$. So the relativistic *kinetic energy* is

$$\frac{m}{\sqrt{1 - v^2}} - m$$

In an inelastic collision, the total kinetic energy before the collision is not the same as the total kinetic energy after the collision. Momentum is conserved, but not kinetic energy. In such a case, there is a loss of kinetic energy, and the mass increases by the amount that the energy decreases.

In the white train frame, the kinetic energy of the double lump after the collision is zero. The kinetic energy of one of the lumps of putty before the collision is

$$\frac{1}{\sqrt{1 - (\frac{1}{2})^2}} - 1 = \frac{2}{\sqrt{3}} - 1$$

The kinetic energy of the other lump of putty is the same, so the total kinetic energy before the collision—which is the kinetic energy lost in the collision—is $4/\sqrt{3} - 2 \approx .309$, the same as the increase in the total mass.

In the black train frame, the kinetic energy before the collision (the kinetic energy of the lump of putty moving with velocity $4/5$) is

$$\frac{1}{\sqrt{1 - (\frac{4}{5})^2}} - 1 = \frac{5}{3} - 1 = \frac{2}{3}$$

The kinetic energy of the double lump after the collision (assuming it has mass M) is

$$\frac{M}{\sqrt{1 - (\frac{1}{2})^2}} - M = \frac{2}{\sqrt{3}}M - M$$

Since, to conserve momentum in all frames, $M = 4/\sqrt{3}$, the loss in kinetic energy in the collision is $4/\sqrt{3} - 2$, as before.

In classical mechanics, momentum is conserved in elastic and inelastic collisions, but kinetic energy is only conserved in elastic collisions—kinetic energy is gained or lost in inelastic collisions. Mass is conserved in all collisions.

Relativistic momentum is conserved in elastic and inelastic collisions, as in classical mechanics. Total relativistic energy is conserved (the mass part and the purely kinetic part) in all collisions, but not the purely kinetic part on its own, because this is used up in increasing the mass part of the total energy in inelastic collisions. Mass is conserved in elastic collisions, where there is no loss or gain in kinetic energy, but not in inelastic collisions, where it gets increased or decreased by kinetic energy.

Page 126:

> You punch "140 grams" and "half the speed
> of light" into a kinetic energy calculator
> and get a really big number.

The velocity of light in vacuum is 299,792,458 meters
per second. Half that is 149,896,229 meters per second. A .140 kilogram mass going at that speed has a
relativistic kinetic energy in joules of

$$\frac{.140\,c^2}{\sqrt{1 - (\frac{1}{2})^2}} - .140\,c^2 \approx 1.9465 \times 10^{15}$$

(A joule is the amount of energy required to increase
the velocity of a 1 kilogram mass by 1 meter per second every second through a distance of 1 meter.) The
energy released by the atomic bomb dropped on Hiroshima was about 15 kilotons of TNT, equivalent to
6.276×10^{13} joules. That makes the energy of the moving mass about 31 times the energy of the Hiroshima
bomb.

Page 131:

> The final results appear almost simple ... the
> years of anxious searching in the dark, with
> their intense longing, their alternations of
> confidence and exhaustion and the final emer-
> gence into the light—only those who have
> experienced it can understand that.

The quotation is from Einstein, 1954a, pp. 289–290.
Our translation combines elements from Bargmann's
translation and the original Glasgow version (p. 11).

Bonus Feature, Pages 133–141

Page 137:

> Since the black train cars are shorter, at
> only 86.6% the length of yours, your chick
> ends up .577 of a train car away from the
> black-train you.

Suppose the black train is going at half the speed of
light to the left relative to the white train, so the velocity is $v = -1/2$. The Lorentz contraction factor is
$\sqrt{1 - v^2} = \sqrt{3}/2 \approx .866$. The eggs hatch at the zero
position on both trains. After 1 second has passed on
the white train, this zero position has moved to the
position

$$\frac{0 - (-\frac{1}{2})}{\frac{\sqrt{3}}{2}} = \frac{1}{\sqrt{3}} \approx .577$$

in the black train reference frame. That's .577 of the
way across the trailing black train car, which is moving
to the left of the white train zero point.

Page 137:

> To be precise, going at half the speed of
> light the white train only gets this far after
> 1.154 black-train seconds.

Because of time dilation, after one second on the white
train, which is moving with half the speed of light relative to the black train, $1/\sqrt{1 - v^2} = 2/\sqrt{3} \approx 1.154$
seconds has passed in the black train reference frame.

Pages 138:

> Then, exactly 1.154 seconds after your egg
> box opened, the black-train you, now .577
> of the way across your train car, freeze-
> frames the thought experiment to deliver

an adorable freshly hatched black chick to
join your remaining white now-chicken .154
of a second old on the white train.

After 1 second on the black train, when the chicks
hatch and one black chick is sent to the white train,
$2/\sqrt{3} \approx 1.154$ seconds have passed on the white train.
The zero position on the black train where the chicks
hatched has moved to the position $-.577$ on the white
train; i.e., $-.577$ of the way across the trailing white-
train car, which is moving to the right of the black
train's zero point. When the black chick shows up at
this position, there is a mature chicken at the zero
position on the white train—the chick that hatched
.154 seconds ago and was not sent to the black train,
and that matured into a chicken after .15 seconds.

Bibliography

Einstein, Albert.1923. "On the electrodynamics of moving bodies." In *The Principle of Relativity*, by H.A. Lorentz, A. Einstein, H. Minkowski, and H. Weyl, with notes by A. Sommerfeld, translated by W. Perrett and G.B. Jeffery. London: Methuen and Company, pp. 37–65. First published in 1905 as "Zur Elektrodynamik bewegter Körper" in *Annalen der Physik*, 17:891–921.

Einstein, Albert. 1954a. "Notes on the origin of the general theory of relativity." In *Ideas and Opinions*, translated by Sonja Bargmann. New York: Bonanza Books, pages 285–290. First published in 1933 as *The Origins of the General Theory of Relativity, Being the First Lecture on the George A. Gibson Foundation in the University of Glasgow Delivered on June 20th, 1933* (Glasgow: Jackson, Wylie, and Company).

Einstein, Albert. 1954b. "What is the theory of relativity?" In *Ideas and Opinions*, translated by Sonja Bargmann. New York: Bonanza Books, pages 227–232. First published in *The London Times*, November 28, 1919.

Mermin, N. David. 2005. *It's About Time: Understanding Einstein's Relativity.* Princeton: Princeton University Press.

Morin, David. 2017. *Special Relativity: For the Enthusiastic Beginner.* CreateSpace.

Shankland, R.S. 1963. "Conversations with Einstein." *Annalen der Physik*, 31:47–57.

Acknowledgments

This book's secret ingredients include: My husband Ian's matter-of-fact confidence that I can actually do all the crazy things I say I'm going to do. Who am I to prove him wrong? My daughter Anouk's insightful critiques served up with a hefty dose of no mercy. You make all my work better. My son Arlo's tirelessly careful readings of the roughest of drafts; he somehow always gets the point, even if it isn't there to get yet. How do you do that? Jeff's wife Robin's ability to keep him sane in the role of devil's advocate by constantly trying to nudge the trains into Minkowski space. My mom (a.k.a #1 fan) Ronit's conviction that whatever I say must be brilliant, even if she has no idea what I'm talking about. Our agent Peter Tallack's openness to taking on another peculiar graphic science book, and to heck with the consequences. Yale editor Jean Black's no-nonsense ability to guide the manuscript through to the finish line, turbulence be damned. Proofreader MaryEllen Oliver's in-the-line-of-duty initiative to set up Legos, in lieu of trains, to see relativity in action for herself. Indexer Kathy Barber's propensity to be intrigued. And finally, our copy editor Mary Pasti's unflinching belief that she could and would understand relativity just by thinking about light, even if she had to point out all the places where the book had to be fixed to make that possible. My hat goes off to you, fearless one!

Index